AF527831

text book of Disaster Management

PREFACE

The major paradigm shift in the field of disaster management started when the courses on Disaster Management was introduced in educational institutions. Teaching of these courses could not become effective due to lack of required book on Disaster Management.

In the present academic set up, the undergraduates and postgraduates rely mostly on the class-room lectures to gain professional knowledge. The blame cannot be put entirely on the student community. Timely non-availability of the required books and more importantly the exorbitant cost of the available books inhibit them to own a book. Supplementation of the class-room knowledge by making available text book of the course curriculum shall certainly stimulate their need for learning and also for better performance in disaster management of livestock sector. It is thus the genesis of this "Text Book of Disaster Management" will propose to be helpful for undergraduate and postgraduate students, Professors, Teachers and Field Veterinarians. It is presumed that the text book will be useful for teachers of Veterinary Colleges engaged in teaching Disaster Management and Veterinarians in Society courses designed by Veterinary Council of India (VCI) and Indian Council of Agricultural Research (ICAR).

Suggestions from the Veterinary Students, Teachers and Field Veterinarians are welcome. Last but not the least, the cooperation and help as well as the interest of Satish Serial Publishing House, Delhi in bringing out this book is highly appreciated.

Nitesh Kumar

Contents

Introduction

DISASTER (French word: "Des" meaning bad or evil and "astre" meaning star)

A serious disruption of the functioning of a society, causing widespread human, material, or environmental losses which exceed the ability of the affected society to cope using its own resources. This disruption occurs when a man-made or natural hazard impacts a vulnerable population.

OR

A serious disruption of the functioning of a community or a society causing widespread human, material, economic or environmental losses which exceed the ability of the affected community or society to cope using its own resources.

OR

A disaster is an event that occurs in most cases suddenly and unexpectedly, causing severe disturbances to people, objects and environment, resulting in loss of life, property and health of the population. Such a situation causes disruption in normal pattern of life and that time there is a need for assistance or immediate outside intervention.

OR

Occurrence of a sudden misfortune which disrupts the normal functioning beyond the capacity to cope is known as disaster.

What is Disaster?

Disruption - to normal patterns of life

Human Effects- loss of life, injury, hardship and adverse effect on health

Effects on Social Structure - destruction of or damage to structures, buildings, communications & essential services

Community needs- emergency shelter, food, clothing, medical, social care etc.

OR

"Disaster" term has its origin from French word "Desastre" which is the combination of two terms **'des'** meaning bad or evil and **'astre'** meaning "star". The combined expression is "Bad or Evil stars". Disaster Management Act, 2005 defines Disaster as: 'Disaster' means a catastrophe, mishap, calamity or grave occurrence in any area arising from natural or manmade causes or by accident or negligence which results in substantial loss of life or human suffering or damage to and destruction of property or damage to or degradation of environment and is of such a nature or magnitude as to be beyond the coping capacity of the community of the affected area.

Disasters in India

- 1769-1770: Great Bengal famine, a third of population perished
- 1783-1901: Chalisa famine to Rajputana famine.
- 1943: Again the Bengal famine
- 1965, 1972, 1979, 1987, 2002, 2005, 2007: the drought years

Major Disasters since 1980

- Bhopal chemical accident, 1984 (3,800 people died)
- Latur earthquake, 1993 (22,000 people died)
- Orissa super cyclone, 1999 (7600 people died)
- Gujarat earthquake, 2001 (20,000 people died)
- Tsunami, 2004 (10,700 people died)
- Mumbai monsoon, 2005 (1000 people died)

India's Vulnerability to Disasters

- 57% land is vulnerable to earthquakes.
- 12% of these is vulnerable to severe earthquakes.
- 68 to 70% land is vulnerable to droughts.
- 12% land is vulnerable to floods.
- 8% land is vulnerable to cyclones.

- Apart from above some cities are vulnerable to chemical and manmade disasters.
- A very long coast line of 5700 Kms is exposed to cyclones.
- 85% of the land area is vulnerable to number of natural hazards
- 22 states are prone to multi hazards.

Scale of Disaster is dependent on

- Lead time available.
- Intensity of hazard.
- Duration.
- Spatial extent.
- Density of population & assets.
- Time of occurrence.
- Vulnerabilities existing in the elements at risk.
- Disaster = Hazard X Vulnerability

Elements at Risk

- People
- Livestock
- Rural housing stock
- Houses vulnerable
- Crops, trees, telephone, electric poles
- Boats, looms, working implements
- Personal property
- Electricity, water and food supplies
- Infrastructure support

THE CRUNCH MODEL

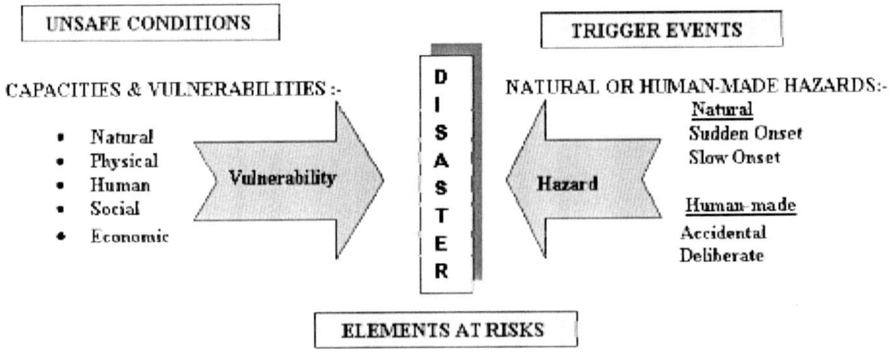

OR

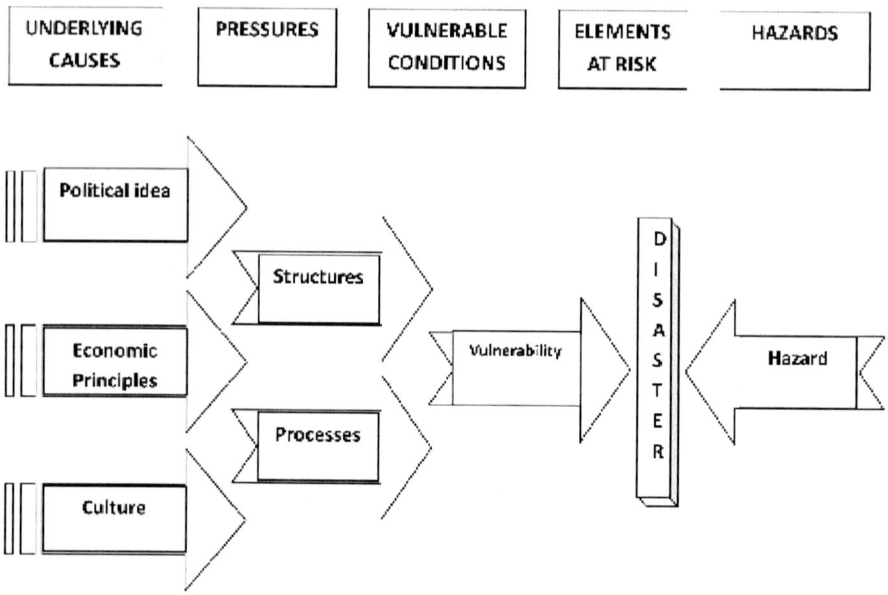

HAZARD: A dangerous condition or event that threatens or has the potential for causing injury to life or damage to property or the environment.

OR

A hazard is a potentially damaging physical event, natural phenomenon or human activity that may cause the loss of life or injury, property damage, social and economic disruption or environmental degradation.

OR

A situation which poses a level of threat to life, health, property and environment is known as hazard.

Origin of Natural Hazards

- Hazards may have different origins: Natural (geological, hydrometeorological and biological) or
- Induced by human processes (environmental degradation and technological hazards).

Categories of Natural Hazards

There are three categories of natural hazards:

Hydro-meteorological, geological and biological.

1. *Hydro-meteorological hazards* are natural processes or phenomena of an atmospheric, hydrological or oceanographic nature.
   - Tropical waves and depressions
   - Tropical Storms
   - Hurricanes
   - Storm surges
   - Torrential rains
   - Floods
   - Tsunamis
   - Droughts
2. *Geological hazards* are natural earth processes or phenomena that include internal earth processes of tectonic origin as well as external processes such as mass movements.
   - Volcanic eruptions
   - Landslides
   - Mud flows
   - Earthquakes
3. *Biological hazards* : Biological weapons (BW) are micro organisms like virus, bacteria and fungi that give rise to disease and when deliberately dispersed in an area can incapicitate, reduce resistance and even cause mortality to humans, domestic animals and crops.

Disease/Agent	Stage	Route of infection	Possible release
Anthrax (*Bacillus anthracis*)	Spores	Skin wounds, Inhalation, ingestion	Spores as aerosol
Plague (*Yersinia pestis*)	Bacteria	Fleas Aerosol	Aerosolization or release of infected fleas
Tularaemia (*Francisella tularensis*)	Bacteria	Aerosol	Aerosolization of the bacteria
Glanders (*Burkholderia mallei*)	Bacteria	Aerosol	Aerosolization of the bacteria
Cholera (*Vibrio cholerae*)	Bacteria	Oral	Contamination of food water sources
Small Pox (*Variolo major*)	Virus	Aerosol	Aerosolization of virus

Vulnerability

A set of conditions and processes resulting from physical, social, economic, and environmental factors, which increase the susceptibility of a community to the impact of hazard.

OR

The condition determined by physical, social, economic and environmental factors or processes, which increase the susceptibility of a community to the impact of hazards.

OR

The extent to which a community, structure, service or geographic areas are likely to be damaged or disrupted by the impact of particular disaster hazards is known as vulnerability.

OR

Vulnerability is the propensity of things to be damaged by a hazard.

OR

The susceptibility to physical or emotional injury or attack is known as vulnerability.

Therefore DISASTER is.....

Disaster = Hazard + Vulnerability

OR

Disaster = Hazard x Vulnerability

OR

$$\text{Disaster Risk} = \frac{\text{Hazard x Vulnerability}}{\text{Capacity}}$$

Capacity: The ability to hold, receive or absorb.

Resilience: It is the capacity of a system, community or society potentially exposed to hazards to adapt, by resisting or changing in order to reach and maintain an acceptable level of functioning and structure.

Risk: The probability of harmful consequences, or expected losses (deaths, injuries, property, livelihoods, economic activity disrupted or environment damage) resulting from interactions between natural or human-induced hazards and vulnerable conditions.

OR

A potential negative impact that may arise from some present process or future event is known as risk.

Risk = Hazard X Vulnerability - Capacity

Acceptable Risk: Refers to the specific value of damage a community is willing to assume.

Type of Disaster

Mainly two type of disaster

1. *Natural Disaster* : e.g. Floods, Droughts, Cyclone, Earthquakes, Landslides, Avalanches, Volcanic eruptions, Heat and cold waves, Climatic changes: Global warming, Sea level rise, Ozone depletion etc.
2. *Man Made Disaster* : e.g. Nuclear disasters, Chemical disasters, Biological disasters, Building fire, Coal fire, Oil fire, Air pollution, Water pollution, Deforestation, Industrial waste water pollution, Road accidents, Rail accidents, Air accidents, Sea accidents etc.

Disasters both manmade and natural have affected mankind since time in memorial

Disasters occur in varied forms

- Some are predictable in advance.
- Some are annual or seasonal.
- Some are sudden and unpredictable.
- Floods -Days and weeks
- Earthquakes -Seconds/minutes
- Cyclones –Days
- Droughts –Months

Grouping	Classification	Hazards
Sub-Group I	Climatic Disasters	Flood, Cyclone, Heat & Cold Wave, Droughts etc.
Sub-Group II	Geological Disasters	Tsunami, Earthquakes, Landslides, Volcanic Eruptions etc.
Sub-Group III	Industrial Disasters	Nuclear, Chemical & Industrial Emergencies.
Sub-Group IV	Accident Related	Fires, Bomb Blasts, Oil Spills, Accidents
Sub-Group V	Biological Disasters	Epidemics, Pest & Cattle Epidemics

Kinds of Natural Disasters

Cyclone

Meteorite

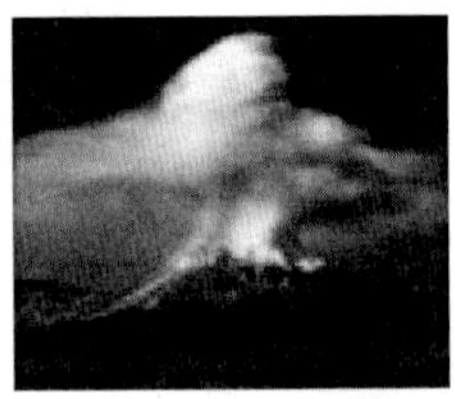

Volcano eruption

Flood

Drought

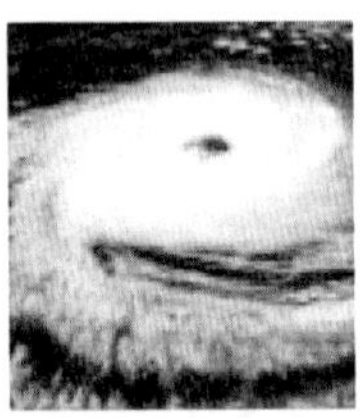

Tornado

Earth quake

Landslide

Avlanche

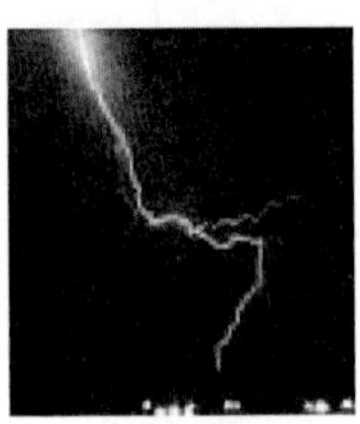

Lightning

Cloud burst

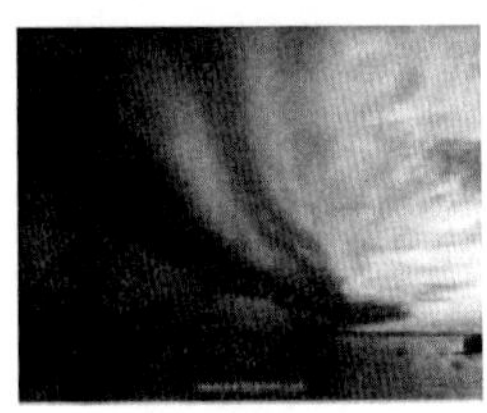

Tsunami

Kinds of Man Made Disasters

CHEMICAL, BIOLOGICAL, RADIOLOGICAL AND NUCLEAR (CBRN)

Road Accidents

Aircraft Crash

General Effects of Disaster

- Loss of life
- Injury
- Damage to and destruction of property
- Damage to infrastructure
- Damage to livelihood source
- National economic loss
- Environmental damage
- Sociological & psychological after effects

Common Terminologies

1. Acceptable risk

The level of potential losses that a society or community considers acceptable given existing social, economic, political, cultural, technical and environmental conditions is known as acceptable risk.

Comment: In engineering terms, acceptable risk is also used to assess and define the structural and non-structural measures that are needed in order to reduce possible harm to people, property, services and systems to a chosen tolerated level, according to codes or "accepted practice" which are based on known probabilities of hazards and other factors.

2. Adaptation

The adjustment in natural or human systems in response to actual or expected climatic stimuli or their effects, which moderates harm or exploits beneficial opportunities is known as adaptation.

Comment: This definition addresses the concerns of climate change and is sourced from the secretariat of the United Nations Framework Convention on Climate Change (UNFCCC). The broader concept of adaptation also applies to non-climatic factors such as soil erosion or surface subsidence. Adaptation can occur in autonomous fashion, for example through market changes, or as a result of intentional adaptation policies and plans. Many disaster risk reduction measures can directly contribute to better adaptation.

3. Biological hazard

Process or phenomenon of organic origin or conveyed by biological vectors, including exposure to pathogenic micro-organisms, toxins and bioactive substances that may cause loss of life, injury, illness or other health impacts, property damage, loss of livelihoods and services, social and economic disruption, or environmental damage.

Comment: Examples of biological hazards include outbreaks of epidemic diseases, plant or animal contagion, insect or other animal plagues and infestations.

4. Building code

A set of ordinances or regulations and associated standards intended to control aspects of the design, construction, materials, alteration and occupancy of structures that are necessary to ensure human safety and welfare, including resistance to collapse and damage.

Comment: Building codes can include both technical and functional standards. They should incorporate the lessons of international experience and should be tailored to national and local circumstances. A systematic regime of enforcement is a critical supporting requirement for effective implementation of building codes.

5. Capacity

The combination of all the strengths, attributes and resources available within a community, society or organization that can be used to achieve agreed goals is known as capacity.

Comment: Capacity may include infrastructure and physical means, institutions, societal coping abilities, as well as human knowledge, skills and collective attributes such as social relationships, leadership and management. Capacity also may be described as capability. Capacity assessment is a term for the process by which the capacity of a group is reviewed against desired goals, and the capacity gaps are identified for further action.

6. Capacity development

The process by which people, organizations and society systematically stimulate and develop their capacities over time to achieve social and economic goals, including through improvement of knowledge, skills, systems, and institutions.

Comment: Capacity development is a concept that extends the term of capacity building to encompass all aspects of creating and sustaining capacity growth over time. It involves learning and various types of training, but also continuous efforts to develop institutions, political awareness, financial resources, technology systems and the wider social and cultural enabling environment.

7. Climate change

(a) The Inter-governmental Panel on Climate Change (IPCC) defines climate change as: "a change in the state of the climate that can be identified (e.g., by using statistical tests) by changes in the mean and/or the variability of its properties, and that persists for an extended period, typically decades or longer. Climate change may be due to natural internal processes or external forcings, or to persistent anthropogenic changes in the composition of the atmosphere or in land use".

(b) The United Nations Framework Convention on Climate Change (UNFCCC) defines climate change as "a change of climate which is attributed directly or indirectly to human activity that alters the composition of the global atmosphere and which is in addition to natural climate variability observed over comparable time periods".

Comment: For disaster risk reduction purposes, either of these definitions may be suitable, depending on the particular context. The UNFCCC definition is the more restricted one as it excludes climate changes attributable to natural causes. The IPCC definition can be paraphrased for popular communications as "A change in the climate that persists for decades or longer, arising from either natural causes or human activity".

8. Contingency planning

A management process that analyses specific potential events or emerging situations that might threaten society or the environment and establishes arrangements in advance to enable timely, effective and appropriate responses to such events and situations.

Comment: Contingency planning results in organized and coordinated courses of action with clearly-identified institutional roles and resources, information processes, and operational arrangements for specific actors at times of need. Based on scenarios of possible emergency conditions or disaster events, it allows key actors to envision, anticipate and solve problems that can arise during crises. Contingency planning is an important part of overall preparedness. Contingency plans need to be regularly updated and exercised.

9. Coping capacity

The ability of people, organizations and systems, using available skills and resources, to face and manage adverse conditions, emergencies or disasters is known as coping capacity.

Comment: The capacity to cope requires continuing awareness, resources and good management, both in normal times as well as during crises or adverse conditions. Coping capacities contribute to the reduction of disaster risks.

10. Corrective disaster risk management

Management activities that address and seek to correct or reduce disaster risks which are already present.

Comment: This concept aims to distinguish between the risks that are already present, and which need to be managed and reduced now, and the prospective risks that may develop in future if risk reduction policies are not put in place.

11. Disaster

A serious disruption of the functioning of a community or a society involving widespread human, material, economic or environmental losses and impacts, which exceeds the ability of the affected community or society to cope using its own resources.

Comment: Disasters are often described as a result of the combination of: the exposure to a hazard; the conditions of vulnerability that are present; and insufficient capacity or measures to reduce or cope with the potential negative consequences. Disaster impacts may include loss of life, injury, disease and other negative effects on human physical, mental and social well-being, together with damage to property, destruction of assets, loss of services, social and economic disruption and environmental degradation.

12. Disaster risk

The potential disaster losses, in lives, health status, livelihoods, assets and services, which could occur to a particular community or a society over some specified future time period.

Comment: The definition of disaster risk reflects the concept of disasters as the outcome of continuously present conditions of risk. Disaster risk comprises different types of potential losses which are often difficult to quantify. Nevertheless, with knowledge of the prevailing hazards and the patterns of population and socio-economic development, disaster risks can be assessed and mapped, in broad terms at least.

13. Disaster risk management

The systematic process of using administrative directives, organizations, and operational skills and capacities to implement strategies, policies and improved coping capacities in order to lessen the adverse impacts of hazards and the possibility of disaster is known as disaster risk management.

Comment: This term is an extension of the more general term "risk management" to address the specific issue of disaster risks. Disaster risk management aims to avoid, lessen or transfer the adverse effects of hazards through activities and measures for prevention, mitigation and preparedness.

14. Disaster risk reduction

The concept and practice of reducing disaster risks through systematic efforts to analyse and manage the causal factors of disasters, including through reduced exposure to hazards, lessened vulnerability of people and property, wise management of land and the environment, and improved preparedness for adverse events.

Comment: A comprehensive approach to reduce disaster risks is set out in the United Nations-endorsed Hyogo Framework for Action, adopted in 2005, whose expected outcome is "The substantial reduction of disaster losses, in lives and the social, economic and environmental assets of communities and countries." The International Strategy for Disaster Reduction (ISDR) system provides a vehicle for cooperation among Governments, organisations and civil society actors to assist in the implementation of the Framework. Note that while the term "disaster reduction" is sometimes used, the term "disaster risk reduction" provides a better recognition of the ongoing nature of disaster risks and the ongoing potential to reduce these risks.

15. Disaster risk reduction plan

A document prepared by an authority, sector, organization or enterprise that sets out goals and specific objectives for reducing disaster risks together with related actions to accomplish these objectives.

Comment: Disaster risk reduction plans should be guided by the Hyogo Framework and considered and coordinated within relevant development plans, resource allocations and programme activities. National level plans needs to be specific to each level of administrative responsibility and adapted to the different social and geographical circumstances that are present. The time frame and responsibilities for implementation and the sources of funding should be specified in the plan. Linkages to climate change adaptation plans should be made where possible.

16. Early warning system

The set of capacities needed to generate and disseminate timely and meaningful warning information to enable individuals, communities and organizations threatened by a hazard to prepare and to act appropriately and in sufficient time to reduce the possibility of harm or loss.

Comment: This definition encompasses the range of factors necessary to achieve effective responses to warnings. A people-centred early warning system necessarily comprises four key elements: knowledge of the risks; monitoring, analysis and forecasting of the hazards; communication or dissemination of alerts and warnings; and local capabilities to respond to the warnings received. The expression "end-to-end warning system" is also used to emphasize that warning systems need to span all steps from hazard detection through to community response.

17. Ecosystem services

The benefits that people and communities obtain from ecosystems are known as ecosystem services.

Comment: This definition is drawn from the Millennium Ecosystem Assessment. The benefits that ecosystems can provide include "regulating services" such as regulation of floods, drought, land degradation and disease, along with "provisioning services" such as food and water, "supporting services" such as soil formation and nutrient cycling, and "cultural services" such as recreational, spiritual, religious and other non-material benefits. Integrated management of land, water and living resources that promotes conservation and sustainable use provide the basis for maintaining ecosystem services, including those that contribute to reduced disaster risks.

18. El Nino-southern oscillation

A complex interaction of the tropical Pacific Ocean and the global atmosphere that results in irregularly occurring episodes of changed ocean and weather patterns in many parts of the world, often with significant impacts over many months, such as altered marine habitats, rainfall changes, floods, droughts, and changes in storm patterns.

Comment: The El Nino part of the El Nino-Southern Oscillation (ENSO) phenomenon refers to the well-above-average ocean temperatures that occur along the coasts of Ecuador, Peru and northern

Chile and across the eastern equatorial Pacific Ocean, while La Nina part refers to the opposite circumstances when well-below-average ocean temperatures occur. The Southern Oscillation refers to the accompanying changes in the global air pressure patterns that are associated with the changed weather patterns experienced in different parts of the world.

19. Emergency management

The organization and management of resources and responsibilities for addressing all aspects of emergencies, in particular preparedness, response and initial recovery steps

Comment: A crisis or emergency is a threatening condition that requires urgent action. Effective emergency action can avoid the escalation of an event into a disaster. Emergency management involves plans and institutional arrangements to engage and guide the efforts of government, non-government, voluntary and private agencies in comprehensive and coordinated ways to respond to the entire spectrum of emergency needs. The expression "disaster management" is sometimes used instead of emergency management.

20. Emergency services

The set of specialized agencies that have specific responsibilities and objectives in serving and protecting people and property in emergency situations is known as emergency services.

Comment: Emergency services include agencies such as civil protection authorities, police, fire, ambulance, paramedic and emergency medicine services, Red Cross and Red Crescent societies, and specialized emergency units of electricity, transportation, communications and other related services organizations.

21. Environmental degradation

The reduction of the capacity of the environment to meet social and ecological objectives and needs is known as environment degradation.

Comment: Degradation of the environment can alter the frequency and intensity of natural hazards and increase the vulnerability of communities. The types of human-induced degradation are varied and include land misuse, soil erosion and loss, desertification, wildland fires, loss of biodiversity, deforestation, mangrove destruction, land, water and air pollution, climate change, sea level rise and ozone depletion.

22. Environmental impact assessment

Process by which the environmental consequences of a proposed project or programme are evaluated, undertaken as an integral part of planning and decision-making processes with a view to limiting or reducing the adverse impacts of the project or programme.

Comment: Environmental impact assessment is a policy tool that provides evidence and analysis of environmental impacts of activities from conception to decision-making. It is utilized extensively in national programming and project approval processes and for international development assistance projects. Environmental impact assessments should include detailed risk assessments and provide alternatives, solutions or options to deal with identified problems.

23. Exposure

People, property, systems, or other elements present in hazard zones that are thereby subject to potential losses.

Comment: Measures of exposure can include the number of people or types of assets in an area. These can be combined with the specific vulnerability of the exposed elements to any particular hazard to estimate the quantitative risks associated with that hazard in the area of interest.

24. Extensive risk

The widespread risk associated with the exposure of dispersed populations to repeated or persistent hazard conditions of low or moderate intensity, often of a highly localized nature, which can lead to debilitating cumulative disaster impacts.

Comment: Extensive risk is mainly a characteristic of rural areas and urban margins where communities are exposed to, and vulnerable to, recurring localised floods, landslides storms or drought. Extensive risk is often associated with poverty, urbanization and environmental degradation. See also "Intensive risk".

25. Forecast

Definite statement or statistical estimate of the likely occurrence of a future event or conditions for a specific area is known as forecast.

Comment: In meteorology a forecast refers to a future condition, whereas a warning refers to a potentially dangerous future condition.

26. Geological hazard

Geological process or phenomenon that may cause loss of life, injury or other health impacts, property damage, loss of livelihoods and services, social and economic disruption, or environmental damage.

Comment: Geological hazards include internal earth processes, such as earthquakes, volcanic activity and emissions, and related geophysical processes such as mass movements, landslides, rockslides, surface collapses, and debris or mud flows. Hydrometeorological factors are important contributors to some of these processes. Tsunamis are difficult to categorize; although they are triggered by undersea earthquakes and other geological events, they are essentially an oceanic process that is manifested as a coastal water-related hazard.

27. Greenhouse gases

Gaseous constituents of the atmosphere, both natural and anthropogenic, that absorb and emit radiation of thermal infrared radiation emitted by the Earth?s surface, the atmosphere itself, and by clouds.

Comment: This is the definition of the Inter-governmental Panel on Climate Change (IPCC). The main greenhouse gases (GHG) are water vapour, carbon dioxide, nitrous oxide, methane and ozone.

28. Hazard

A dangerous phenomenon, substance, human activity or condition that may cause loss of life, injury or other health impacts, property damage, loss of livelihoods and services, social and economic disruption, or environmental damage.

Comment: The hazards of concern to disaster risk reduction as stated in footnote 3 of the Hyogo Framework are "hazards of natural origin and related environmental and technological hazards and risks." Such hazards arise from a variety of geological, meteorological, hydrological, oceanic, biological, and technological sources, sometimes acting in combination. In technical settings, hazards are described quantitatively by the likely frequency of occurrence of different intensities for different areas, as determined from historical data or scientific analysis. See other hazard-related terms in the terminology:

Biological hazard; Geological hazard; Hydrometeorological hazard; Natural hazard; Socio-natural hazard; Technological hazard.

29. Hydrometeorological hazard

Process or phenomenon of atmospheric, hydrological or oceanographic nature that may cause loss of life, injury or other health impacts, property damage, loss of livelihoods and services, social and economic disruption, or environmental damage.

Comment: Hydrometeorological hazards include tropical cyclones (also known as typhoons and hurricanes), thunderstorms, hailstorms, tornados, blizzards, heavy snowfall, avalanches, coastal storm surges, floods including flash floods, drought, heatwaves and cold spells. Hydrometeorological conditions also can be a factor in other hazards such as landslides, wildland fires, locust plagues, epidemics, and in the transport and dispersal of toxic substances and volcanic eruption material.

30. Intensive risk

The risk associated with the exposure of large concentrations of people and economic activities to intense hazard events, which can lead to potentially catastrophic disaster impacts involving high mortality and asset loss.

Comment: Intensive risk is mainly a characteristic of large cities or densely populated areas that are not only exposed to intense hazards such as strong earthquakes, active volcanoes, heavy floods, tsunamis, or major storms but also have high levels of vulnerability to these hazards. See also "Extensive risk".

31. Land-use planning

The process undertaken by public authorities to identify, evaluate and decide on different options for the use of land, including consideration of long term economic, social and environmental objectives and the implications for different communities and interest groups, and the subsequent formulation and promulgation of plans that describe the permitted or acceptable uses.

Comment: Land-use planning is an important contributor to sustainable development. It involves studies and mapping; analysis of economic, environmental and hazard data; formulation of alternative land-use decisions; and design of long-range plans for different geographical and administrative scales. Land-use planning can help to mitigate disasters and reduce risks by discouraging settlements and construction of key installations in hazard-prone areas, including consideration of service routes for transport, power, water, sewage and other critical facilities.

32. Mitigation

The lessening or limitation of the adverse impacts of hazards and related disasters is known as mitigation.

Comment: The adverse impacts of hazards often cannot be prevented fully, but their scale or severity can be substantially lessened by various strategies and actions. Mitigation measures encompass engineering techniques and hazard-resistant construction as well as improved environmental policies and public awareness. It should be noted that in climate change policy, "mitigation" is defined differently, being the term used for the reduction of greenhouse gas emissions that are the source of climate change.

33. National platform for disaster risk reduction

A generic term for national mechanisms for coordination and policy guidance on disaster risk reduction that are multi-sectoral and inter-disciplinary in nature, with public, private and civil society participation involving all concerned entities within a country is known as national platform for disaster risk reduction.

Comment: This definition is derived from footnote 10 of the Hyogo Framework. Disaster risk reduction requires the knowledge, capacities and inputs of a wide range of sectors and organisations, including United Nations agencies present at the national level, as appropriate. Most sectors are affected directly or indirectly by disasters and many have specific responsibilities that impinge upon disaster risks. National platforms provide a means to enhance national action to reduce disaster risks, and they represent the national mechanism for the International Strategy for Disaster Reduction.

34. Natural hazard

Natural process or phenomenon that may cause loss of life, injury or other health impacts, property damage, loss of livelihoods and services, social and economic disruption, or environmental damage.

Comment: Natural hazards are a sub-set of all hazards. The term is used to describe actual hazard events as well as the latent hazard conditions that may give rise to future events. Natural hazard events can be characterized by their magnitude or intensity, speed of onset, duration, and area of extent. For example, earthquakes have short durations and usually affect a relatively small region, whereas droughts are slow to develop and fade away and often affect large regions. In some cases hazards may be coupled, as in the flood caused by a hurricane or the tsunami that is created by an earthquake.

35. Preparedness

The knowledge and capacities developed by governments, professional response and recovery organizations, communities and individuals to effectively anticipate, respond to, and recover from, the impacts of likely, imminent or current hazard events or conditions.

Comment: Preparedness action is carried out within the context of disaster risk management and aims to build the capacities needed to efficiently manage all types of emergencies and achieve orderly transitions from response through to sustained recovery. Preparedness is based on a sound analysis of disaster risks and good linkages with early warning systems, and includes such activities as contingency planning, stockpiling of equipment and supplies, the development of arrangements for coordination, evacuation and public information, and associated training and field exercises. These must be supported by formal institutional, legal and budgetary capacities. The related term "readiness" describes the ability to quickly and appropriately respond when required.

36. Prevention

The outright avoidance of adverse impacts of hazards and related disasters is known as prevention.

Comment: Prevention (i.e. disaster prevention) expresses the concept and intention to completely avoid potential adverse impacts through action taken in advance. Examples include dams or embankments that eliminate flood risks, land-use regulations that do not permit any settlement in high risk zones, and seismic engineering designs that ensure the survival and function of a critical building in any likely earthquake. Very often the complete avoidance of losses is not feasible and the task transforms to that of mitigation. Partly for this reason, the terms prevention and mitigation are sometimes used interchangeably in casual use.

37. Prospective disaster risk management

Management activities that address and seek to avoid the development of new or increased disaster risks is known as prospective disaster risk management.

Comment: This concept focuses on addressing risks that may develop in future if risk reduction policies are not put in place, rather than on the risks that are already present and which can be managed and reduced now. See also Corrective disaster risk management.

38. Public awareness

The extent of common knowledge about disaster risks, the factors that lead to disasters and the actions that can be taken individually and collectively to reduce exposure and vulnerability to hazards.

Comment: Public awareness is a key factor in effective disaster risk reduction. Its development is pursued, for example, through the development and dissemination of information through media and educational channels, the establishment of information centres, networks, and community or participation actions, and advocacy by senior public officials and community leaders.

39. Recovery

The restoration, and improvement where appropriate, of facilities, livelihoods and living conditions of disaster-affected communities, including efforts to reduce disaster risk factors is known as recovery.

Comment: The recovery task of rehabilitation and reconstruction begins soon after the emergency phase has ended, and should be based on pre-existing strategies and policies that facilitate clear institutional responsibilities for recovery action and enable public participation. Recovery programmes, coupled with the heightened public awareness and engagement after a disaster, afford a valuable opportunity to develop and implement disaster risk reduction measures and to apply the "build back better" principle.

40. Residual risk

The risk that remains in unmanaged form, even when effective disaster risk reduction measures are in place, and for which emergency response and recovery capacities must be maintained.

Comment: The presence of residual risk implies a continuing need to develop and support effective capacities for emergency services, preparedness, response and recovery together with socio-economic policies such as safety nets and risk transfer mechanisms.

41. Resilience

The ability of a system, community or society exposed to hazards to resist, absorb, accommodate to and recover from the effects of a hazard in a timely and efficient manner, including through the preservation and restoration of its essential basic structures and functions.

Comment: Resilience means the ability to "resile from" or "spring back from" a shock. The resilience of a community in respect to potential hazard events is determined by the degree to which the community has the necessary resources and is capable of organizing itself both prior to and during times of need.

42. Response

The provision of emergency services and public assistance during or immediately after a disaster in order to save lives reduces health impacts, ensure public safety and meet the basic subsistence needs of the people affected.

Comment: Disaster response is predominantly focused on immediate and short-term needs and is sometimes called "disaster relief". The division between this response stage and the subsequent recovery stage is not clear-cut. Some response actions, such as the supply of temporary housing and water supplies, may extend well into the recovery stage.

43. Retrofitting

Reinforcement or upgrading of existing structures to become more resistant and resilient to the damaging effects of hazards is known as retrofitting.

Comment: Retrofitting requires consideration of the design and function of the structure, the stresses that the structure may be subject to from particular hazards or hazard scenarios, and the practicality and costs of different retrofitting options. Examples of retrofitting include adding bracing to stiffen walls, reinforcing pillars, adding steel ties between walls and roofs, installing shutters on windows, and improving the protection of important facilities and equipment.

44. Risk

The combination of the probability of an event and its negative consequences is known as risk.

Comment: This definition closely follows the definition of the ISO/ IEC Guide 73. The word "risk" has two distinctive connotations: in popular usage the emphasis is usually placed on the concept of chance or possibility, such as in "the risk of an accident"; whereas in technical settings the emphasis is usually placed on the consequences, in terms of "potential losses" for some particular cause, place and period. It can be noted that people do not necessarily share the same perceptions of the significance and underlying causes of different risks. See other risk-related terms in the terminology: Acceptable risk; Corrective disaster risk management; Disaster risk; Disaster risk management; Disaster risk reduction; Disaster risk reduction plans; Extensive risk; Intensive risk; Prospective disaster risk management; Residual risk; Risk assessment; Risk management; Risk transfer.

45. Risk assessment

A methodology to determine the nature and extent of risk by analysing potential hazards and evaluating existing conditions of vulnerability that together could potentially harm exposed people, property, services, livelihoods and the environment on which they depend.

Comment: Risk assessments (and associated risk mapping) include: a review of the technical characteristics of hazards such as their location, intensity, frequency and probability; the analysis of exposure and vulnerability including the physical social, health, economic and environmental dimensions; and the evaluation of the effectiveness of prevailing and alternative coping capacities in respect to likely risk scenarios. This series of activities is sometimes known as a risk analysis process.

46. Risk management

The systematic approach and practice of managing uncertainty to minimize potential harm and loss is known as risk management.

Comment: Risk management comprises risk assessment and analysis, and the implementation of strategies and specific actions to control, reduce and transfer risks. It is widely practiced by organizations to minimise risk in investment decisions and to address operational risks such as those of business disruption, production failure, environmental damage, social impacts and damage from fire and natural hazards. Risk management is a core issue for sectors such as water supply, energy and agriculture whose production is directly affected by extremes of weather and climate.

47. Risk transfer

The process of formally or informally shifting the financial consequences of particular risks from one party to another whereby a household, community, enterprise or state authority will obtain resources from the other party after a disaster occurs, in exchange for ongoing or compensatory social or financial benefits provided to that other party.

Comment: Insurance is a well-known form of risk transfer, where coverage of a risk is obtained from an insurer in exchange for ongoing premiums paid to the insurer. Risk transfer can occur informally within family and community networks where there are reciprocal expectations of mutual aid by means of gifts or credit, as well as formally where governments, insurers, multi-lateral banks and other large risk-bearing entities establish mechanisms to help cope with losses in major events. Such mechanisms include insurance and re-insurance contracts, catastrophe bonds, contingent credit facilities and reserve funds, where the costs are covered by premiums, investor contributions, interest rates and past savings, respectively.

48. Socio-natural hazard

The phenomenon of increased occurrence of certain geophysical and hydrometeorological hazard events, such as landslides, flooding, land subsidence and drought, that arise from the interaction of natural hazards with overexploited or degraded land and environmental resources.

Comment: This term is used for the circumstances where human activity is increasing the occurrence of certain hazards beyond their natural probabilities. Evidence points to a growing disaster burden from such hazards. Socio-natural hazards can be reduced and avoided through wise management of land and environmental resources.

49. Structural and non-structural measures

Structural measures: Any physical construction to reduce or avoid possible impacts of hazards, or application of engineering techniques to achieve hazard-resistance and resilience in structures or systems; Non-structural measures: Any measure not involving physical construction that uses knowledge, practice or agreement to reduce risks and impacts, in particular through policies and laws, public awareness raising, training and education.

Comment: Common structural measures for disaster risk reduction include dams, flood levies, ocean wave barriers, earthquake-resistant construction, and evacuation shelters. Common non-structural measures include building codes, land use planning laws and their enforcement, research and assessment, information resources, and public awareness programmes. Note that in civil and structural engineering, the term "structural" is used in a more restricted sense to mean just the load-bearing structure, with other parts such as wall cladding and interior fittings being termed non-structural.

50. Sustainable development

Development that meets the needs of the present without compromising the ability of future generations to meet their own needs is known as sustainable development.

Comment: This definition coined by the 1987 Brundtland Commission is very succinct but it leaves unanswered many questions regarding the meaning of the word development and the social, economic and environmental processes involved. Disaster risk is associated with unsustainable elements of development such as environmental degradation, while conversely disaster risk reduction can contribute to the achievement of sustainable development, through reduced losses and improved development practices.

51. Technological hazards

A hazard originating from technological or industrial conditions, including accidents, dangerous procedures, infrastructure failures or specific human activities, that may cause loss of life, injury, illness or other health impacts, property damage, loss of livelihoods and services, social and economic disruption, or environmental damage.

Comment: Examples of technological hazards include industrial pollution, nuclear radiation, toxic wastes, dam failures, transport accidents, factory explosions, fires, and chemical spills. Technological hazards also may arise directly as a result of the impacts of a natural hazard event.

52. Vulnerability

The characteristics and circumstances of a community, system or asset that make it susceptible to the damaging effects of a hazard is known as vulnerability.

Comment: There are many aspects of vulnerability, arising from various physical, social, economic, and environmental factors. Examples may include poor design and construction of buildings, inadequate protection of assets, lack of public information and awareness, limited official recognition of risks and preparedness measures, and disregard for wise environmental management. Vulnerability varies significantly within a community and over time. This definition identifies vulnerability as a characteristic of the element of interest (community, system or asset) which is independent of its exposure. However, in common use the word is often used more broadly to include the element's exposure.

❑❑❑

Disaster Management

Disaster Management

The operational activities required to prepare, mitigate and respond to disasters.

OR

"Disaster management" means a continuous and integrated process of planning, organising, coordinating and implementing measures which are necessary or expedient for :

- Prevention of danger or threat of any disaster.
- Mitigation or reduction of risk of any disaster or its severity or consequences.
- Capacity-building.
- Preparedness to deal with any disaster.
- Prompt response to any threatening disaster situation or disaster.
- Assessing the severity or magnitude of effects of any disaster.
- Evacuation, rescue and relief.
- Rehabilitation and reconstruction.

Aims of Disaster Management

- Reduce (Avoid, if possible) the potential losses from hazards.
- Assure prompt and appropriate assistance to victims when necessary.
- Achieve rapid and durable recovery.

Disaster Risk Management

Components	Activities
Risk Identification	Hazard and Vulnerability Studies
Risk Reduction	Prevention and Mitigation
Risk Transfer	Insurance / Risk Financing
Preparedness, Alert, Response	Forecasting, dissemination of warnings, implementing preparedness measures
Recovery	Rehabilitation and Reconstruction

Risk Identification: Risk identification is the systematic use of available information to determine the probability of certain adverse events occurring and the related consequences.

Risk Identification - examples

- Hazard data collection and mapping -identification, dimensioning, frequency of the hazard.
- Vulnerability assessments -identification of who (population) and what (infrastructure) are vulnerable to the hazard.
- Risk assessments -determining the expected loss from a hazardous event.
- Post-disaster assessments.

Risk Reduction: Risk reduction is an evolving area of disaster risk management aimed at risk elimination or reduction by intervening in the vulnerability. In other words, risk reduction involves clear and explicit effort to avoid the occurrence of disasters. Risk reduction comprises two components: Prevention and Mitigation.

OR

Disaster Risk Reduction : The framework of elements considered with the possibilities to minimize vulnerabilities and disaster risks throughout a community, to avoid (prevention) or to limit (mitigation and preparedness) the adverse impacts of hazards, within the broad context of sustainable development.

OR

Risk reduction is an evolving area of disaster risk management aimed at risk elimination or reduction by intervening in the vulnerability of a community, of an area, of infrastructure, etc.

OR

The conceptual framework of elements considered with the possibilities to minimize vulnerabilities and disaster risks throughout a society, to avoid (prevention) or to limit (mitigation and preparedness) the adverse impacts of hazards, within the broad context of sustainable development is known as disaster risk reduction.

Risk Reduction -examples

- Physical measures
- Socio-economic measures
- Environmental measures
- Post-disaster measures

Disaster Risk Reduction (DRR) aims to reduce the damage caused by natural hazards like earthquakes, floods, droughts and cyclones, through an ethic of prevention. There is no such thing as a 'natural' disaster, only natural hazards. Disasters often follow natural hazards. A disaster's severity depends on how much impact a hazard has on society and the environment. The scale of the impact in turn depends on the choices we make for our lives and for our environment. These choices relate to how we grow our food, where and how we build our homes, what kind of government we have, how our financial system works and even what we teach in schools. Each decision and action makes us more vulnerable to disasters - or more resilient to them.

DRR is defined by UN agency UNDP as: "The conceptual framework of elements considered with the possibilities to minimize vulnerabilities and disaster risks throughout a society, to avoid (prevention) or to limit (mitigation and preparedness) the adverse impacts of hazards, within the broad context of sustainable development."

Disaster risk reduction includes disciplines like disaster management, disaster mitigation and disaster preparedness, but DRR is also part of sustainable development. In order for development activities to be sustainable they must also reduce disaster risk. On the other hand, unsound development policies will increase disaster risk -

and disaster losses. Thus, DRR involves every part of society, every part of government, and every part of the professional and private sector. The more governments, UN agencies, organizations, businesses and civil society understand risk and vulnerability, the better equipped they will be to mitigate disasters when they strike and save more lives.

The scope of Disaster Risk Reduction is much broader and deeper than conventional emergency management. It is a systematic approach to identifying, assessing and reducing the risks of disaster. It aims to reduce socio-economic vulnerabilities to disaster as well as dealing with the environmental and other hazards that trigger them: here it has been strongly influenced by the mass of research on vulnerability that has appeared in print since the mid-1970s. It is the responsibility of development and relief agencies alike and it should be an integral part of the way such organisations do their work, not an add-on or one-off action. DRR is very wide-ranging, therefore. There is potential for DRR initiatives in just about every sector of development and humanitarian work.

The evolution of disaster management thinking and practice since the 1970s has seen a progressively wider and deeper understanding of why disasters happen, accompanied by more integrated, holistic approaches to reduce their impact on society. The modern paradigm of disaster management - disaster risk reduction (DRR) - represents the latest step along this path. DRR is a relatively new concept in formal terms, but it embraces much earlier thinking and practice, and it is now being widely embraced by international agencies, governments, disaster planners and civil society organisations.

DRR is such an all-embracing concept that it has proved difficult to define or explain in detail, although the broad idea is clear enough. Inevitably, there are different definitions of the term in the technical literature but it is generally understood to mean the broad development and application of policies, strategies and practices to minimise vulnerabilities and disaster risks throughout society. The term 'disaster risk management' (DRM) is often used in the same context and to mean much the same thing: a systematic approach to identifying, assessing and reducing risks of all kinds associated with hazards and human activities. It is more properly applied to the operational aspects of DRR: the practical implementation of DRR initiatives.

There have been growing calls for greater clarity about the components of DRR, and about indicators of progress towards resilience - a challenge which the international community took up at the UN's World Conference on Disaster Reduction (WCDR) in Kobe, Japan, in 2005, only days after the 2004 Indian Ocean earthquake. The WCDR began the process of pushing international agencies and national governments beyond the vague rhetoric of most policy statements and towards setting clear targets and commitments for DRR. The first step in this process was the formal approval at the WCDR of the Hyogo Framework for Action (2005-2015) (HFA). This is the first internationally accepted framework for DRR. It sets out an ordered sequence of objectives (outcome - strategic goals - priorities), with five priorities for action attempting to 'capture' the main areas of DRR intervention. The UN's biennial Global Platform for disaster risk reduction provides an opportunity for the UN and its member states to review progress against the Hyogo Framework. It held its first session from 5-7 June 2007 in Geneva, Switzerland.

United Nation (UN) initiatives have helped to refine and promote the concept at international level, stimulated initially by the UN's designation of the 1990s as the International Decade for Natural Disaster Reduction.

Some issues and challenges in DRR

Priorities

It is unrealistic to expect progress in every aspect of DRR: capacities and resources are insufficient. Governments and other organisations have to make what are in effect 'investment decisions', choosing which aspects of DRR to invest in, when, and in what sequence. This is made more complicated by the fact that many of the interventions advocated are developmental rather than directly related to disaster management. Most existing DRR guidance sidesteps this issue. One way of focusing is to consider only actions that are intended specifically to reduce disaster risk. This would at least distinguish from more general efforts towards sustainable development. The concept of 'invulnerable development' attempts this: in this formulation, invulnerable development is development directed towards reducing vulnerability to disaster, comprising decisions and activities that are intentionally designed and implemented to reduce risk and susceptibility, and also raise resistance and resilience to disaster.

Partnerships and inter-organisational co-ordination

No single group or organisation can address every aspect of DRR. DRR thinking sees disasters as complex problems demanding a collective response. Co-ordination even in conventional emergency management is difficult, for many different organisations may converge on a disaster area to assist. Across the broader spectrum of DRR, the relationships between different types of organisation and between different sectors (public, private and non-profit, as well as communities) become much more extensive and complex. DRR requires strong vertical and horizontal linkages (central-local relations become important). In terms of involving civil society organisations, it should mean thinking broadly about which types of organisation to involve (i.e. not just conventional NGOs but also such organisations as trades unions, religious institutions, amateur radio operators (as in the USA and India), universities and research institutions).

Communities and their organisations

Traditional emergency management/civil defence thinking makes two misleading assumptions about communities. First, it sees other forms of social organisation (voluntary and community-based organisations, informal social groupings and families) as irrelevant to emergency action. Spontaneous actions by affected communities or groups (e.g. search and rescue) are viewed as irrelevant or disruptive, because they are not controlled by the authorities. The second assumption is that disasters produce passive 'victims' who are overwhelmed by crisis, or dysfunctional behaviour (panic, looting, self-seeking activities). They therefore need to be told what to do, and their behaviour must be controlled - in extreme cases, through the imposition of martial law. There is plenty of sociological research to refute such 'myths'.

An alternative viewpoint, informed by a considerable volume of research, emphasises the importance of communities and local organisations in disaster risk management. The rationale for community-based disaster risk management that it responds to local problems and needs, capitalises on local knowledge and expertise, is cost-effective, improves the likelihood of sustainability through genuine ownership of projects, strengthens community technical and organisational capacities, and empowers people by enabling them to tackle these and other challenges. Local people and organisations are the main actors in risk reduction and disaster response in any case.

Governance

The DRR approach requires redefining the role of government disaster reduction. It is generally agreed that national governments should be main actors in DRR: they have a duty to ensure the safety of citizens, the resources and capacity to implement large-scale DRR, a mandate to direct or co-ordinate the work of others, and they create the necessary policy and legislative frameworks. These policies and programmes have to be coherent. More research is needed into why some governments are more successful than others in disaster management. There is still no general consensus on what drives changes in policy and practice. The shifting relationship between central government and other actors is another area requiring research.

Accountability and rights

The principle of accountability lies at the heart of genuine partnership and participation in DRR. It applies to state institutions that are expected to be accountable through the democratic process, and to private sector and non-profit organizations which are not subject to democratic control. Accountability is still an emerging issue in disaster reduction work. Accountability should be primarily towards those who are vulnerable to hazards and affected by them. Many organisations working in international aid and development are now committing themselves to a 'rights-based' approach. This tends to encompass both human rights (i.e. those that are generally accepted through international agreements) and other rights that an agency believes should be accepted as human rights. In such contexts, the language of rights may be used vaguely, with a risk of causing confusion. Security against disasters is not generally regarded as a right although it is addressed in some international codes, usually indirectly. The idea of a 'right to safety' is also being discussed in some circles. A stronger and more sustainable disaster risk reduction and resilience movement world-wide that leads to increased responsibility for reinforcing resilience to disaster:

- A dynamic and trend-setting forum for decision makers, partners, experts and practitioners to announce initiatives, launch products, share information, promote campaigns, and provide evidence around disaster risk reduction.
- Directions and new alliances for the development and use of new tools and methodologies aimed at understanding and

applying the economics and investment in disaster risk reduction; building resilience in communities; and, climate change adaptation.

- A forum to discuss progress and consult over a post-2015 framework for disaster risk reduction (post-HFA), including on issues raised in consultations.
- Events that follow-up and progress on the 2011 Global Platform (e.g. update on disaster loss in schools and hospitals, accounting for disaster losses, the status of National Platforms, and the Children's Charter for Disaster Risk Reduction).

Disaster Risk Reduction Measures: There are five categories of Disaster Risk Reduction measures:

1. Physical
2. Socio-economic
3. Environmental
4. Management and institutional and
5. Post disaster.

1. Physical Measures

Physical Risk Reduction Measures

- Physical risk reduction measures are structural and non-structural measures that are undertaken to help avoid or alleviate the damage and losses that result from hazards.

Structural Measures

- Structural measures refer to any physical construction to reduce or avoid possible impacts of hazards, which include engineering measures and construction of hazard-resistant and protective structures and infrastructure.

Examples

- Construction;
- Strengthening of facilities;
- Altering of the environment.

Non-structural Measures

- Non-structural measures refer to policies, awareness, knowledge development, public commitment, and methods and operating practices, including participatory mechanisms and the provision of information, which can reduce risk and related impacts.

Examples

- Engineering design such as designing for vibration, lateral loads, wind loads, load surcharges, flood resistance;
- Physical development plans that identify land use zones that are developable and those that are hazard prone;
- Development regulations (design standards, building codes).

2. Socio-economic Measures

- Socio-economic risk reduction measures are those that are designed to address gaps and weaknesses in the systems whereby communities and society as a whole prepare to respond to disaster events.

Examples

- Build awareness of hazards and vulnerabilities at the community level;
- Create community mutual assistance networks;
- Develop employment protection programs.

3. Environmental Measures

Environmental Risk Reduction measures

- Environmental risk reduction measures are those that are designed to protect existing, or rehabilitate degraded, environmental systems that have the capacity to reduce the impact of natural hazards.

Examples

- Policies and programmes such as development control or environmental impact assessments that reduce or eliminate the effect of human activities on the environment;

- Physical measures that restore or fortify damaged environmental systems such as mangrove replanting, coral reef protection, reforestation of critical watersheds or restoration of degraded river courses.

4. Management and Institutional Measures

Institutionalising disaster risk reduction requires:

- Education, training and professional competence, access to information and political will.
- The professional training of engineers, planners, economists, social scientists and other managers to include hazards and disaster risk reduction within their normal area of competence.
- Information is a critical element in planning for disaster risk reduction.

5. Post-disaster Measures

- Reconstruction and rehabilitation work during the post-disaster period determines the capability of the system to weather future hazards.
- The recovery period can be seen as a window of opportunity for implementing a comprehensive disaster risk reduction programme.

Active and Passive Risk Reduction Measures

Active Risk Reduction Measures

- Active Measures are those used by the authorities to promote desired actions by offering **incentives**. These measures are often associated with development programmes in areas of low income.

Examples of incentives are

- Tax exemptions and other incentives for those who fulfill planning regulations such as adoption of the building code;
- Economic assistance (grants and preferential loans);
- Subsidies on safety equipment, safer building materials, etc;
- Provision of financial opportunities for the construction of key facilities: safer buildings, shelters, refuge points, storage;

- Training and education;
- Public information dissemination and awareness raising;
- Promotion of voluntary insurance; and
- Creation of community organizations.

Advantages of Active RR measures

- May produce better results because they:
- Tend to promote a safety culture.
- Do not rely on the economic capacity of the community.
- Do not rely on enforcement by local authorities.

Disadvantage Active DRR measures

- More costly -require large budgets, skilled workforce and extensive administration.

Passive Risk Reduction measures

- Passive measures are those used by the authorities to prevent undesired actions, as for example **controls and penalties** (as opposed to incentives). However, these actions are usually more appropriate for well-established local authorities in areas with higher incomes.

Examples

- Land use regulations;
- Requirements to conform with building design codes;
- Checking compliance of controls on site;
- Not approving the construction of utilities and infrastructure in areas where development is undesired;
- Imposing court proceedings, fines, closure orders on offenders; and Compulsory insurance.

Disaster Risk Management

- The systematic management of
  - administrative decisions,
  - organization,
  - operational skills and
  - capacities to implement

- policies,
- strategies and coping capacities of the society and communities to lessen the impacts of natural hazards and related environmental and technological disasters.
- This comprises all forms of activities, including structural and non-structural measures to avoid (prevention) or to limit (mitigation and preparedness) adverse effects of hazards.

Prevention : Activities to provide outright avoidance of the adverse impact of hazards and means to minimize related environmental, technological and biological disasters.

Mitigation : Structural and non-structural measures undertaken to limit the adverse impact of natural hazards, environmental degradation and technological hazards. It embraces all measures taken to reduce both the effects of the hazard itself and the vulnerable conditions to it in order to reduce the scale of a future disaster. Therefore mitigation activities can be focused on the hazard itself or the elements exposed to the threat.

OR

The process of reducing of the severity to risks is known as mitigation.

Preventable hazards and impacts that can be mitigated

Preventable Hazards

- Oil spill
- Toxic waste contamination
- Industrial pollution
- Landslide
- Flooding

Hazards which can be mitigated

- Oil spill
- Toxic waste contamination
- Industrial pollution
- Landslide
- Flooding

Hazards whose impacts can be mitigated

- Earthquake
- Hurricane
- Volcanic eruption
- Landslide
- Flooding

Risk Transfer: Risk Transfer refers to instruments that share/hedge economic risks before losses occur.

Risk Transfer - examples

- Budget self-insurance;
- Market Insurance and Reinsurance;
- Public asset coverage;
- Risk pooling and diversification;
- Risk financing.

Budget Self-insurance: Budget self-insurance is the measure taken by individual property owners to allocate a modest yearly budget for improvement, maintenance and selected retrofit investments for their property (buildings, land, infrastructure) which helps to reduce future expected losses in the event of a disaster.

Market Insurance: Market insurance is the mechanism by which losses are stabilised through pre-payment in the form of regular premium payments.

Public asset coverage: Public asset coverage financing is the process by which governments seek to finance the reconstruction of damaged critical public assets. Usually this comes from annual budgets, but needs to be augmented by outside sources.

Risk pooling & diversification: Diversification of the risk pool is done by banking with others from other geographic areas or industries which can help reduce insurance premiums for all participants.

Catastrophe Bond: A cat bond is a bond that is purchased by investors which transfers risks to the global capital market.

Risk transfer ensures that funds are readily available to rectify the damage or replace the facility, should a loss occur. It does not reduce actual vulnerability. As a result, risk transfer mechanisms should always be implemented in support of, rather than as a replacement for, broad disaster risk reduction initiatives, such as improved building practices and land use planning.

DISASTER MANAGEMENT CYCLE

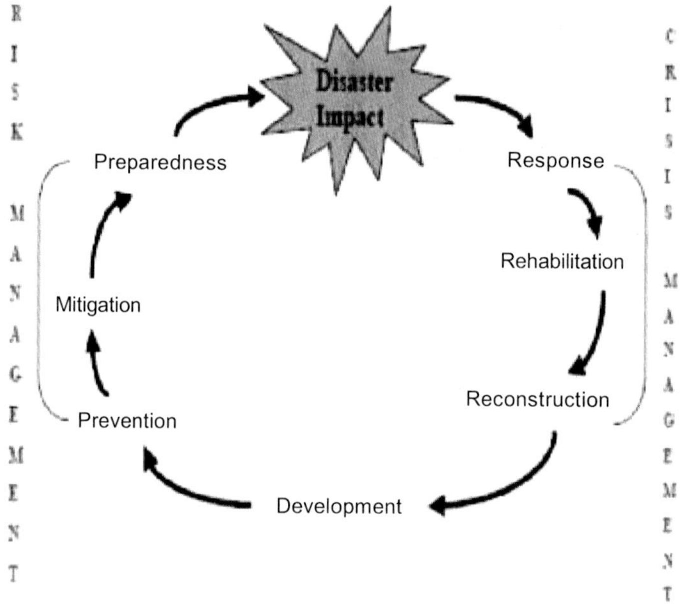

Preparedness

It is a protective process which embraces measures to enable governments, communities and individuals to respond rapidly to disaster situations to cope with them effectively. Preparedness includes the formation of viable emergency plans, development of warning systems, the maintenance of inventories and the training of personnel.

OR

Activities and measures taken in advance to ensure effective response to the impact of hazards, including the issuance of timely and effective early warnings and the temporary evacuation of people and property from threatened locations.

OR

The state of being ready with adequate knowledge, skill and material resources to cope with disaster is known as preparedness.

Disaster Preparedness

Disaster preparedness aims at minimizing the adverse effects of a hazard:

- Through effective precautionary actions.
- Ensure timely, appropriate and efficient organization and delivery of emergency response following the impact of a disaster.
- Vulnerability analysis and mapping to include resources.
- Assess strengthening requirements and execute.
- Funding for preparedness must be arranged.
- People's cooperation through political leaders, elders, volunteers and NGOs.
- Create lead time by interpreting warnings.
- Plan to include movement of resources with time frame.
- Aim to reduce the destructive potential of cyclones, timely & appropriate relief to victims and quick & durable recovery.

Disaster preparedness framework

Component of Preparedness		
Vulnerability Assessment	**Planning**	**Institutional framework**
Information system	Resource base	Warning systems
Response mechanisms	Public education and training	Rehearsals

Alert: A formal declaration of the near or imminent occurence of a hazard. This information should lead emergency organisations to activate previously established mechanisms and the population to take specific precautions.

Response: The provision of assistance or intervention during or immediately after a disaster to meet the life preservation and basic subsistence needs of those people affected. It can be of an immediate, short-term, or protracted duration.

Disaster Response Activities

- Warning;
- Evacuation/Mitigation;

- Search and Rescue;
- Assessment;
- Emergency Relief;
- Logistics and Supply;
- Communication and information Management;
- Survivor Response and coping;
- Security;
- Coordination;
- Expedite rehabilitation and reconstruction.

Recovery-Rehabilitation and Reconstruction

Recovery: Decisions and actions taken after a disaster with a view to restoring or improving the pre-disaster living conditions of the stricken "community", while encouraging and facilitating necessary adjustments to reduce disaster risk.

Rehabilitation: This is the transition period that begins during response to re-establish temporary critical basic services in the short-term.

Reconstruction: Reconstruction is the process of infrastructure repair, restoration of the production system and resumption of the population's normal life pattern.

Phases of Disaster Risk Management

1. Pre-Disaster

- Risk Identification;
- Risk Reduction;
- Prevention and Mitigation;
- Risk Transfer;
- Preparedness.

2. Post-Disaster

- Response;
- Recovery -Rehabilitation and Reconstruction.

Damage from natural hazards

1. *Direct Damage:* Direct damage is all damage sustained by immovable assets and inventories of finished and semi-finished products, raw materials, other materials and spare parts.
2. *Indirect Damage*: This is damage to the flows of goods that cease to be produced or the services that cease to be provided during a period of time beginning almost immediately after the disaster and possibly extending into the rehabilitation and reconstruction phase.

Disaster Risk Management and Development

- Natural events such as earthquakes, volcanoes, hurricanes and floods, within the context of expanding population and economic growth, pose a growing threat to national and regional development strategies.
- Natural hazards can delay future development due to the loss of resources. They may also cause a shift of scarce resources earmarked for development projects to emergency response and reconstruction following the events.
- They can also depress the investment climate. Proactive disaster risk management, which involves priority investment in mitigation and prevention, must be an essential strategy for sustainable development in the Caribbean Region.

RELEASE MODEL

Action

Advocacy Education Awareness Local National International	**Advocacy Development** Local National International	**Development Mitigation Preparedness** Local National	**Developmental Relief** Local	**Hazard Reduction** Local National International

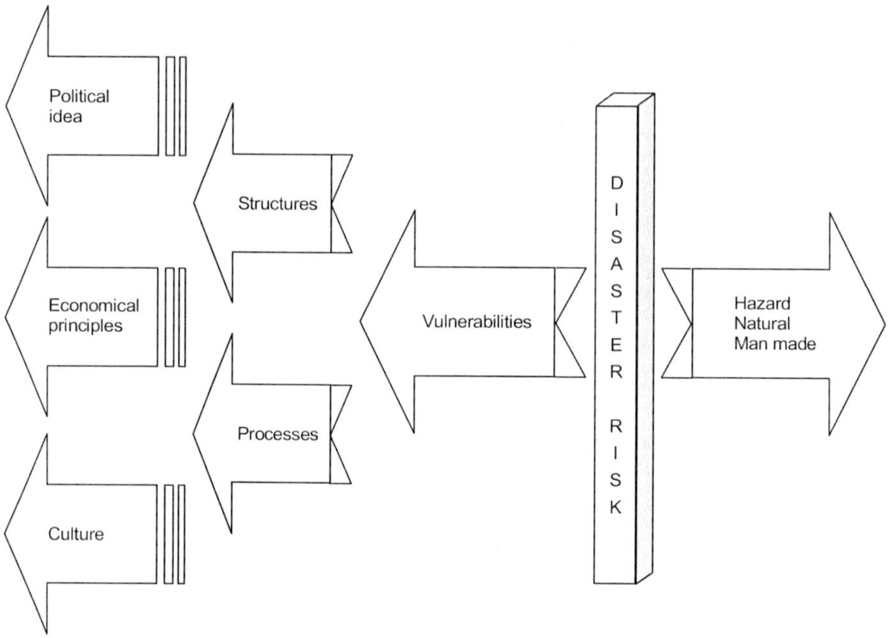

OUTCOME

Underlying causes addressed	Pressures released	Safe conditions	Protected elements	Reduced hazards

International Mandate

The five priorities for action of the Hyogo Framework for Action (HFA)

- Ensure that Disaster Risk Reduction (DRR) is a national and a local priority with a strong institutional basis for implementation.
- Identify, assess and monitor disaster risks and enhance early warning.
- Use knowledge, innovation and education to build a culture of safety and resilience at all levels.
- Reduce the underlying risk factors.
- Strengthen disaster preparedness for effective response at all levels.

Extreme weather event vulnerable zone

- 5700 km cost line
- 40 m hectare flood prone
- 80% of net sown area is drought prone spreading over 116 districts.
- Sub Himalayan/west Ghats are landslide zone.

INDIA Prone to Multi Hazards

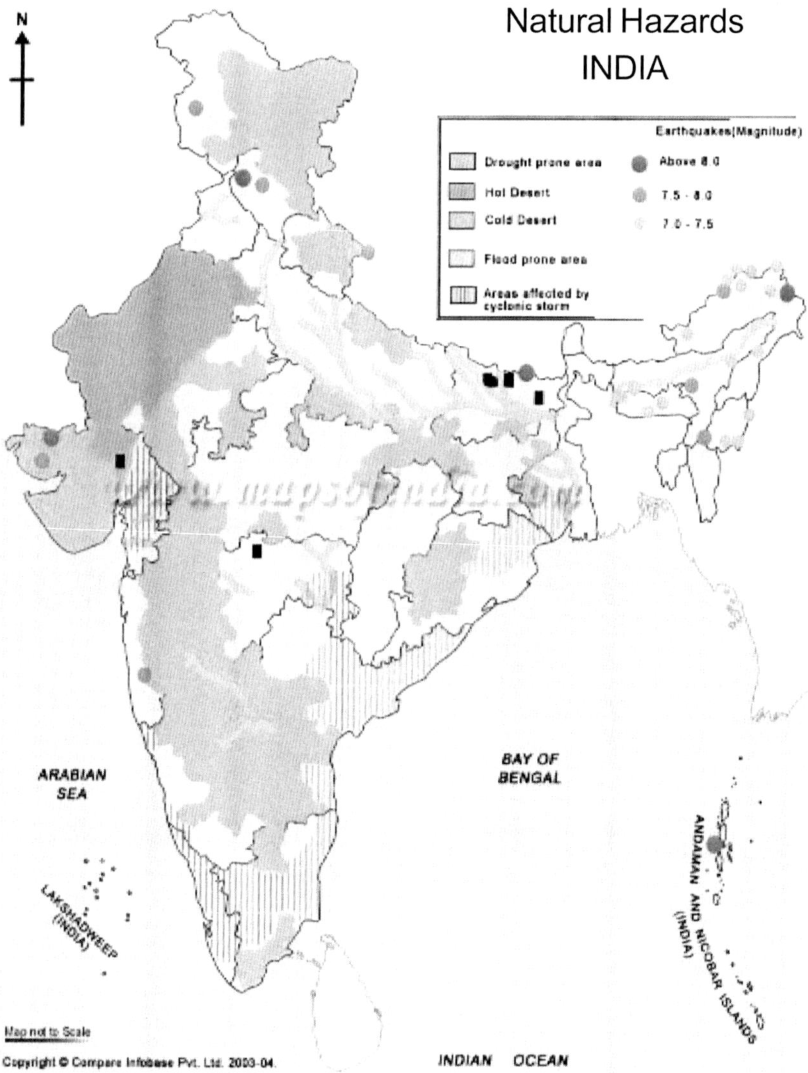

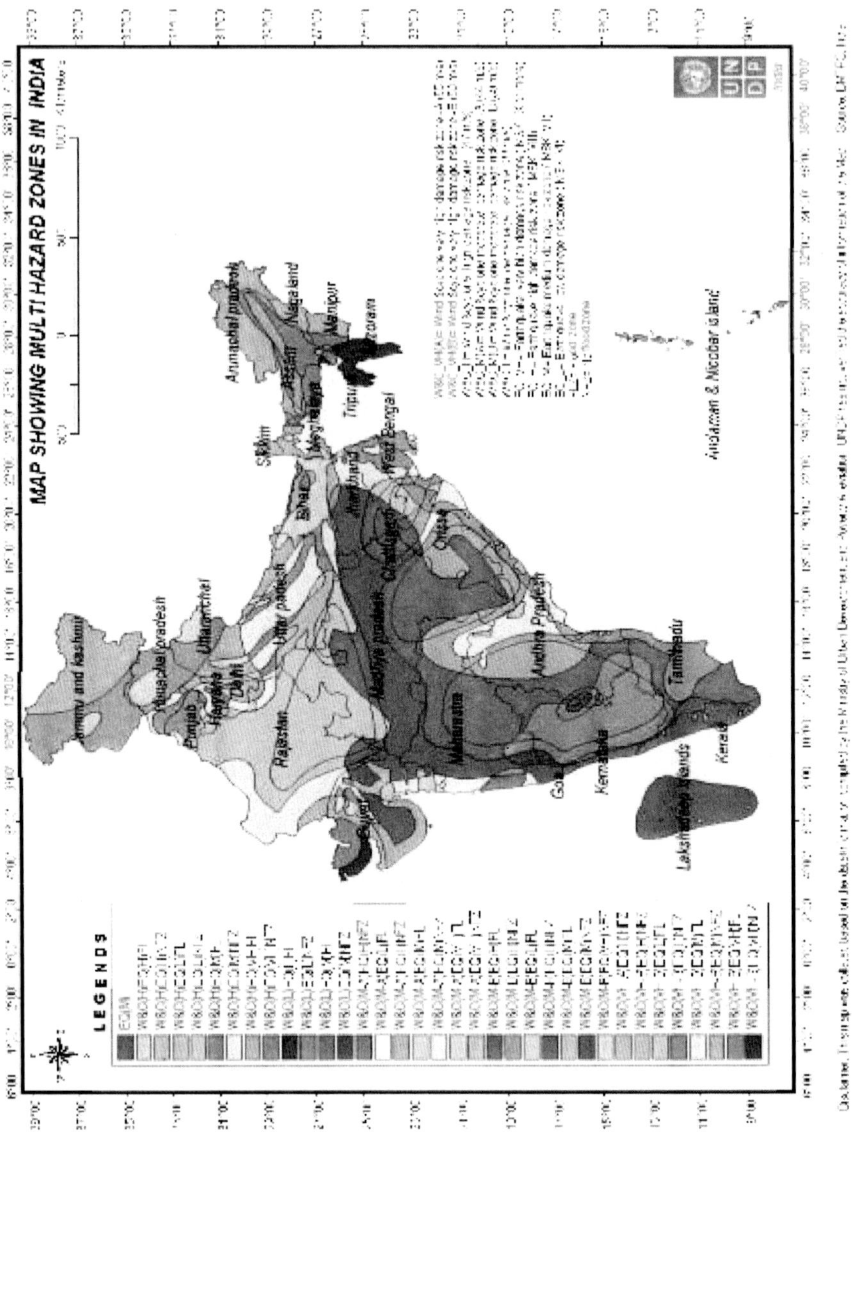
MAP SHOWING MULTI HAZARD ZONES IN INDIA
LEGENDS
Jammu and kashmir
Himachal pradesh
Punjab
Haryana
Delhi
Uttaranchal
Rajastan
Uttar pradesh
Gujarat
Madhya pradesh
Chattisgarh
Bihar
Jharkhand
Sikkim
West Bengal
Meghalaya
Assam
Arunachal pradesh
Nagaland
Manipur
Tripura
Orissa
Maharashtra
Goa
Karnataka
Andhra Pradesh
Tamilnadu
Kerala
Lakshadweep Islands
Andaman & Nicobar island

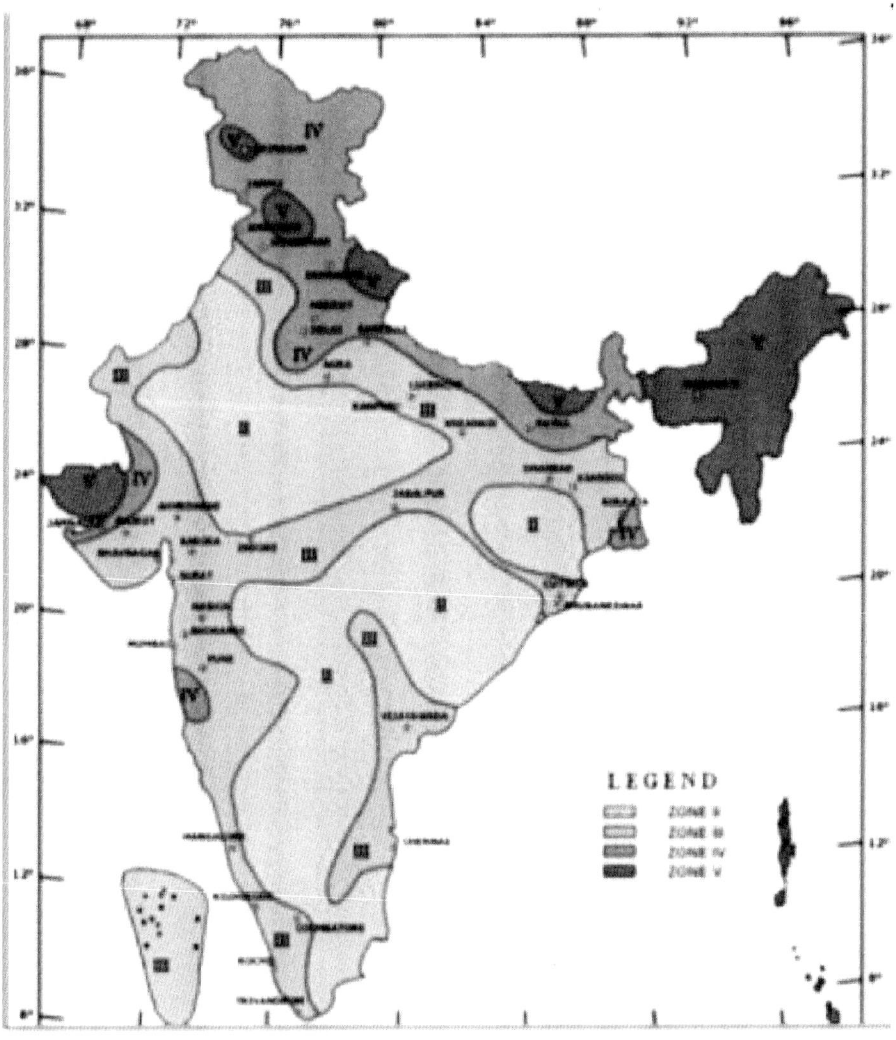

Zone	Magnitude
Zone V	Very High Risk Quakes of Magnitude **8** and greater
Zone IV	High Risk Quakes upto Magnitude **7.9**
Zone III	Moderate Risk Quakes upto Magnitude **6.9**
Zone II	Seismic Disturbances upto Magnitude **4.9**

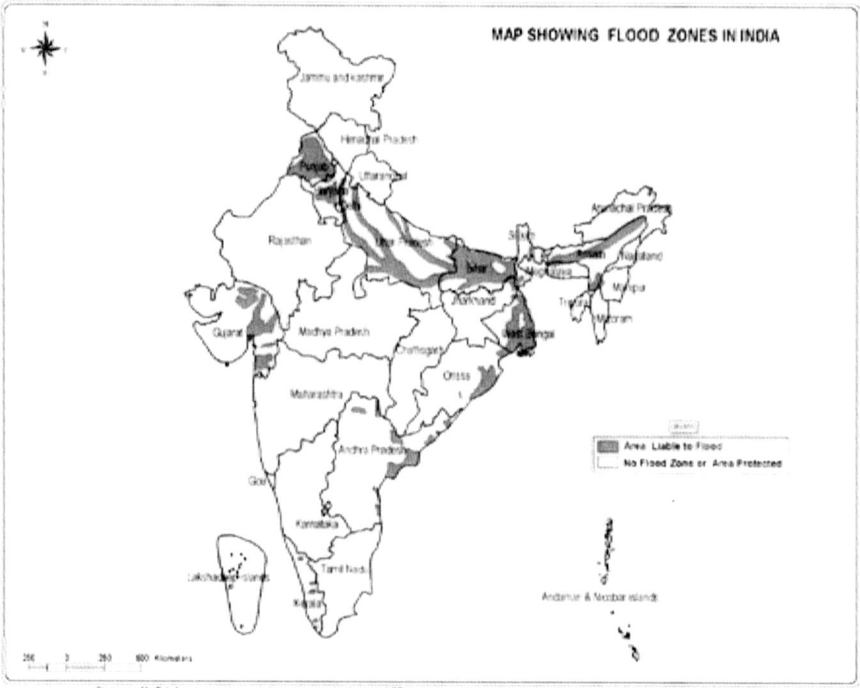
MAP SHOWING FLOOD ZONES IN INDIA
Jammu and Kashmir
Himachal Pradesh
Rajasthan
Gujarat
Madhya Pradesh
Maharashtra
Orissa
Andhra Pradesh
Karnataka
Tamil Nadu
Andaman & Nicobar Islands
Area Liable to Flood
No Flood Zone or Area Protected

MAP SHOWING WIND&CYCLONE ZONES IN INDIA
Himachal Pradesh
Uttar Pradesh
Madhya Pradesh
Maharashtra
Andhra Pradesh
Karnataka
Tamil Nadu
Andaman & Nicobar Islands
Very High Damage Risk Zone-A(Vb= 55m/s)
Very High Damage Risk Zone-B(Vb= 50m/s)
High Damage Risk Zone (Vb= 47m/s)
Moderate Damage Risk Zone-A(Vb= 44m/s)
Moderate Damage Risk Zone-B
Low Damage Risk Zone (Vb= 33m/s)

Tenth Five Year Plan emphasizes on

- Planning for safer national development
- Disaster prevention, mitigation and preparedness measures
- Capacity building
- Training and education at all levels

Major areas of concern are

- Inclusion of disaster management in school curriculum.
- Disaster management, awareness and preparedness in school, college & universities.
- Preparation of the disaster management plans at educational institutions.

Participatory Disaster Risk Assessment (PDRA) - Essential for spatial planning

- The risks associated with disasters and its impacts on the community, and how it compounds the community vulnerability.
- Adaptive capacities of the community to cope and address the risk associated with the natural disasters.
- Enables to the community to development micro level action plan to reduce the risk.

Steps involved in PDRA

- Community understanding the Preparedness concept & process.
- Hazard Mapping: the type and nature of hazard faced by community.
- Vulnerability Assessment: livelihood, livestock, habitat, structure etc, and also analyzes the underlying causes.
- Capacity Assessment: It enables the community to assess the existing capacity at local level and how it can be utilized to reduce the risk.
- DRR Action plan: a community based micro level planning with focus on risk reduction, adaptive capacity and preparedness.

Adaptation-DRR Design

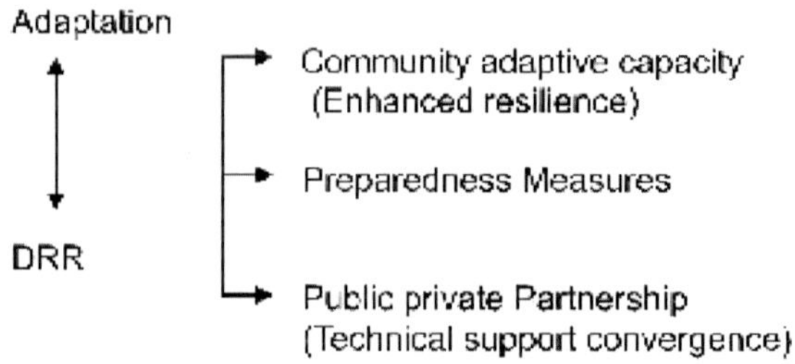

Community capacity

- Disaster management committee (DMC): 50 villages.
- Task Force-warning, rescue & evacuation, first aid, relief management, shelter management & watsan.
- Disaster Risk Reduction Fund.
- Self Help Group: 50 SHGs.
- Crop diversification (Rice, Potato, wheat): 1200 farmers benefitted.

DRR measures

- Structural mitigation measures: 15 Boats, 17 High raised platform, 2 evacuation routes, 2 flood shelters.
- Preparation of village level contingency plan-for local level advocacy.
- Local level Advocacy using RTI-700 applications filed to know the status of various schemes and government implementing activities in the project location. e.g. Bhadas village in Bhadas Panchayat got electricity under the Rural Electrification Program.

Public-private partnership

- Collaboration with ATMA & KVK for technical input - farmers.
- District Disaster Management Plan for Madhubani District in collaboration with Sphere India and SDMA and NDMA India.
- Input to NDMA as its NGO task force member to develop flood management guideline.
- NGO guidelines on humanitarian response.
- International campaign on climate change particularly adaptation.

Sectors

- Shelter;
- Livelihood;
- Health and Nutrition;
- Water/Sanitation;
- Education;
- Forestry/Environment;
- Land use.

Example of Health & Nutrition

Policies & Legislations

- Quality standards & guidelines for infrastructure focus on hazard-resistant construction of hospitals and health facilities/ centres.
- Land use planning technique used for planning for location of new hospitals/ health centres.

Advocacy & Awareness

- Health workers are aware of NDMA Guidelines on medical preparedness and mass casualty management for disasters.
- Hospital Disaster Management Toolkits, Community Health Workers manuals etc are available for health workers to use.

Coordination & Capacity development

- Health training: Anganwadi workers, ASHAs and health volunteers are trained on various aspects of health and hygiene during emergencies.
- Anganwadi centres are strengthened (eg. safe structure, DRM training for workers, more staff during risk prone season etc.).
- Health workers are involved in disaster preparedness efforts (eg. inclusion of disaster management and first aid into ASHA training curriculum, ASHA workers part of Village Disaster Management Teams etc.).
- A contingency plan exists for health services and delivery during the monsoon season (eg. equipping paramedical staff in villages that are hard to reach).

Risk-proofing & Monitoring

- Primary health centres and Community health centres are fully equipped to continue functioning during disasters.

Engagement of local bodies & communities

- Communities are trained in counselling skills to deal with post-disaster trauma.

Disaster Risk Reduction Stakeholders

Local

- State Government;
- Non government Organizations;
- Community based Organizations.

National

- Central Government.

International

- WHO;
- FAO;
- UNICEF, International NGOs.

The Disaster Risk Reduction Framework

- Risk awareness and assessment including hazard analysis and vulnerability/capacity analysis;
- Knowledge development including education, training, research and information;
- Public commitment and institutional frameworks, including organizational, policy, legislation and community action;
- Application of measures including environmental management, land-use and urban planning, protection of critical facilities, application of science and technology, partnership and networking, and financial instruments;
- Early warning systems including forecasting, dissemination of warnings, preparedness measures and reaction capacities.

Importance

- Risk reduction reduces vulnerability;
- Reduced vulnerability reduces impact of future hazardous events;
- Reduced impact means quicker recovery;
- Quicker recovery allows return to development activities.

Vulnerability Assessment

- Vulnerability assessments are systematic examinations of population, buildings, infrastructure and selected geographic areas to identify who, what and where are susceptible to damage from the effects of natural hazards.

Vulnerability Analysis

Determine vulnerable elements

- Determine the elements that are vulnerable to the hazard: population, neighborhoods, or environmental zones;
- Identify critical facilities such as: Shelters, schools, hospitals, clinics and fire stations;
- Lifelines such as water, sewerage, telecommunications, energy and gas systems;
- Housing;
- Historical buildings and archaeological sites.

Determining characteristics of vulnerable population

- number of residents;
- health condition;
- socio-economic characteristics.

Establish community's response capability

- level of knowledge on risk issues;
- resilience shown by the community;
- participation in the design, test and planning execution.

Role Players in Disasters

- *People*: Individuals, House -Holds, Volunteers.
- *Gram Panchayat*: Sarpanch, Panchayati Secretary, Panchayati Members.
- *Village Elders*: Caste/Community/Religious Leaders, Teachers, Doctors, Engineers, Retired Army & Police Personnel.
- *Government Departmental Officers*: Agriculture, Medical, Engineers (Housing, Roads & Buildings, Irrigation), Revenue Department, Public Health, Police etc., NGOs.

Ministries Responsible

- Natural Disaster (Other than Drought) - MHA
- Drought Relief - Agriculture
- Air Accidents - Civil Aviation
- Railway Accidents - Railways
- Chemical Disasters - Environment & Forests
- Bio Disasters - Health
- Nuclear Disasters - DAE
- Use of NBC Agents by Terrorists - MHA

Dynamics of Disasters : There is a high probability of a low probability event happening somewhere sometime soon.......

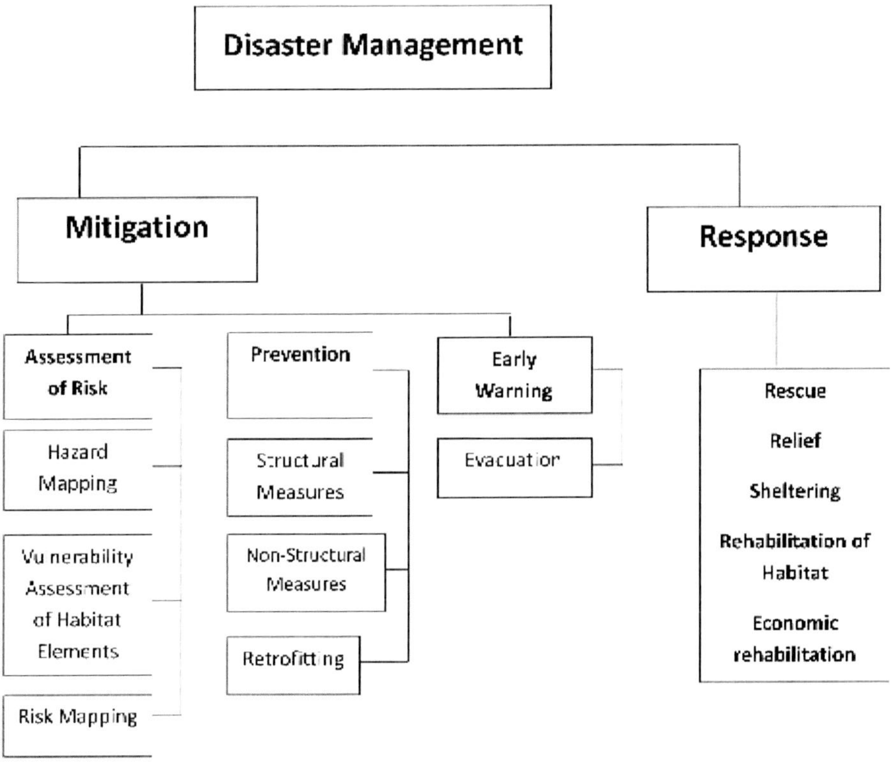

The unpredictability of disaster events and the high risk and vulnerability profiles make it imperative to strengthen disaster preparedness, mitigation and enforcement of guidelines.

❑❑❑

Disaster Management in India

"**DISASTER MANAGEMENT**" means a continuous and integrated process of planning, organising, coordinating and implementing measures which are necessary or expedient for:

- Prevention of danger or threat of any disaster.
- Mitigation or reduction of risk of any disaster or its severity or consequences.
- Capacity-building.
- Preparedness to deal with any disaster.
- Prompt response to any threatening disaster situation or disaster.
- Assessing the severity or magnitude of effects of any disaster.
- Evacuation, rescue and relief.
- Rehabilitation and reconstruction.

Management of Disaster in India

- Paradigm shift in approach.
  - a) From Response and Relief Centre to:
    - i) Prevention and restoration
    - ii) Mitigation and preparedness
  - b) From Ministry of Agriculture to Ministry of Home in 2002.
- High power Committee under J C Pant-1999.
  - i) Culture of preparedness
  - ii) Culture of quick response
  - iii) Culture of strategic thinking
  - iv) Culture of Mitigation.

- All party National Committee under chairmanship of P.M.-2001
- D M Act - 2005
  - i) Constitution of NDMA, SDMA, DDMA
  - ii) Constitution of NDRF / SDRF
  - iii) Provision of Mitigation / Legal Actions
  - iv) Responsibility to each department
- Inclusion in Five year plan

Development cannot be sustainable unless D.M is built into development process

- Recommendations by 13th Finance Commission
- Inclusion of curriculum in Education system
- Community Preparedness / Awareness.

Organizations

- National Disaster Management Aurhority (NDMA)
- State Disaster Management Authority (SDMA)
- District Disaster Management (DDMA)

National Disaster Management Authority (NDMA)

- NDMA constitutes of members not exceeding ten:
  - The Prime Minister of India–Ex Officio Chair Person.
  - Nine other members nominated by Prime Minister.

Role of NDMA

- Lay down policies of Disaster Management.
- Approve the National plan.
- Approve plans prepared by the ministries or departments of the Government of India in accordance with National Plan.
- Lay down guidelines to be followed by the state auth in drawing up State Plan.
- Lay down guidelines to be followed by the different ministries or departments of the Government of India for the purpose of integrating the measures for prevention of disaster or the mitigation of its effects in their development plans and projects.

- Coordinate the enforcement and implementation of the policy and plans for DM.
- Arrange for and oversee the provisions of funds for the purpose of mitigation measures, preparedness and response.
- Provide such support to other countries affected by major disasters as determined by the Central Government.
- Take such other measures, for the preventing of disaster or mitigation or for preparedness and capacity building for dealing with threatening disaster situation or disaster as it may consider necessary.
- Constitute an Advisory Committee consisting of experts in the field of disaster management and having practical experience of disaster management at the national, state or district level, to make recommendations on different aspects of disaster management.

National Executive Committee (NEC)

- The secretary to the Government of India in charge of ministry or department of Government of India, having administrative control of the disaster management, who will be ex officio - "Chair Person".
- The secretaries to the Government of India in the ministries or departments having administrative control of the agriculture, atomic energy, defence, drinking water supply, environment and forests, finance (expenditure), health, power, rural development, science and technology, space, telecommunication, water resources and the Chief of the Integrated Defence staff of the Chiefs of Staff Committee ex officio.

Role of NEC

- Act as the coordinating and monitoring body for DM.
- Prepare national plan to be approved by NDMA.
- Coordinate and monitor the implementation of National Policy.
- Lay down guidelines for preparing DM plans by different ministries or departments of Government of India and state authorities.

- Provide technical assistance to the state government and the state auth for preparing their DM plans in accordance with the guidelines laid down by NDMA.
- Monitor the implementation of the guidelines laid down by the NDMA for integrating of measures for prevention of disasters and mitigation by the ministries or departments in their development plans and projects.
- Monitor the implementation of National Plan and the plans prepared by the ministries or departments of government of India.
- Monitor, coordinate and give directions regarding the mitigations and preparedness measures to be taken by different ministries or departments and agencies of the government.
- Evaluate the preparedness at all government levels for the purpose of responding to any threatening disaster situation or disaster and give directions, where necessary for enhancing such preparedness.
- Plan and coordinate specialized training programme for DM for different levels of officers, employees and voluntary rescue workers.
- Coordinate response in the event of any threatening disaster situation or disaster.
- Lay down guidelines for or give directions to the concerned ministries or departments of Government of India, state government and state auth regarding measures to be taken by them in response to any threatening disaster situation or disaster.
- Require only department or agency of the government to make available to NDMA or state authorities such men or material resources as are available with it for the purposes of emergency response, rescue and relief.
- Advice, assist and coordinate the activities of the ministries or departments of Government of India, state auth, statutory bodies, other government or non government organizations and others engaged in DM.
- Provide necessary technical assistance or give advice to the state auth and district auth for carrying out their functions under this Act.

- Promote general education and awareness in relation to DM.
- Perform such other functions as NDMA may require it to perform.

State Disaster Management Authority (SDMA)

- The Chief Minister of the State, shall be Chairperson, ex officio.
- Nine other members nominated by Chief Minister.
- The Chairperson of the State Executive Committee, ex officio.
- The Chairperson of the State Authority may designate one of the members nominated to be the Vice-Chairperson of the State Authority.

Role of SDMA

- Lay down the State Disaster Management Policy.
- Lay down the State Plan in accordance with the guidelines laid down by the National Authority.
- Approve the Disaster Management Plans prepared by the departments of the Government of the State.
- Coordinate the implementation of the Plan.
- Oversee the provision of funds for mitigation and preparedness measures.
- Review the development plans of the different departments of the State and ensure that prevention and mitigation measures are integrated therein.
- Review the measures being taken for mitigation, capacity building and preparedness by the departments of the Government of the State and issue such guidelines or directions as may be necessary.

State Executive Committee (SEC)

Consists of the following members:

- The Chief Secretary to the State Government, who shall be Chairperson, ex officio.
- Four Secretaries to the Government of the State of such departments as the State Government may think fit, ex officio.

Role of SEC

- Coordinate and monitor the implementation of the National Policy, the National Plan and the State Plan.
- Examine the vulnerability of different parts of the State to different forms of disasters and specify measures to be taken for their prevention or mitigation.
- Lay down guidelines for preparation of disaster management plans by the departments of the Government of the State and the District Authorities.
- Monitor the implementation of Disaster Management Plans prepared by the departments of the Government of the State and District Authorities.
- Lay down guidelines to be followed by the departments of the Government of the State for the purposes of integration of measures for prevention of disasters and mitigation in their development plans and projects and provide necessary technical assistance.
- Evaluate preparedness at all governmental or non-governmental levels to respond to any threatening disaster situation or disaster and give directions, where necessary, for enhancing such preparedness.
- Coordinate response in the event of any threatening disaster situation or disaster.
- Give directions to any Department of the Government of the State or any other authority or body in the State regarding actions to be taken in response to any threatening disaster situation or disaster.
- Promote general education, awareness and community training in regard to the forms of disasters to which different parts of the State are vulnerable and the measures that may be taken by such community to prevent the disaster, mitigate and respond to such disaster.
- Advise, assist and coordinate the activities of the Departments of the Government of the State, District Authorities, statutory bodies and other governmental and non-governmental organisations engaged in disaster management.

- Provide necessary technical assistance or give advice to District Authorities and local authorities for carrying out their functions effectively.
- Advise the State Government regarding all financial matters in relation to disaster management.
- Examine the construction, in any local area in the State and, if it is of the opinion that the standards laid for such construction for the prevention of disaster is not being or has not been followed, may direct the District Authority or the local authority, as the case may be, to take such action as may be necessary to secure compliance of such standards.
- Provide information to the National Authority relating to different aspects of disaster management.
- Lay down, review and update State level response plans and guidelines and ensure that the district level plans are prepared, reviewed and updated.
- Ensure that communication systems are in order and the disaster management drills are carried out periodically.
- Perform such other functions as may be assigned to it by the State Authority.

Powers & Functions of SEC

- Control and restrict, vehicular traffic to, from or within, the vulnerable or affected area.
- Control and restrict the entry of any person into, his movement within and departure from, a vulnerable or affected area.
- Remove debris, conduct search and carry out rescue operations.
- Provide shelter, food, drinking water, essential provisions, healthcare and services in accordance with the standards laid down by the National Authority and State Authority.
- Give direction to the concerned Department of the Government of the State, any District Authority or other authority, within the local limits of the State to take such measure or steps for rescue, evacuation or providing immediate relief saving lives or property, as may be necessary in its opinion.

- Require any department of the Government of the State or any other body or authority or person in charge of any relevant resources to make available the resources for the purposes of emergency response, rescue and relief.
- Require experts and consultants in the field of disasters to provide advise and assistance for rescue and relief.
- Procure exclusive or preferential use of amenities from any authority or person as and when required.
- Construct temporary bridges or other necessary structures and demolish unsafe structures which may be hazardous to public.
- Ensure that non-governmental organisations carry out their activities in an equitable and non-discriminatory manner.
- Disseminate information to public to deal with any threatening disaster situation or disaster.
- Take such steps as the Central Government or the State Government may direct in this regard or take such other steps as are required or warranted by the form of any threatening disaster situation or disaster.

District Disaster Management Authority (DDMA)

- The District Authority will consist of members, not exceeding seven namely:
- Collector or District Magistrate or Deputy Commissioner, as the case may be, of the district who shall be Chairperson, *ex officio.*
- Chief Executive Officer of the District Authority.
- Superintendent of Police, *ex officio.*
- Chief Medical Officer of the district, *ex officio.*
- Three other district level officers, to be appointed by the State Government.

Role of DDMA

- Prepare a Disaster Management plan including district response plan for the district.

- Coordinate and monitor the implementation of the National Policy, State Policy, National Plan, State Plan and District Plan.
- Ensure that the areas in the district vulnerable to disasters are identified and measures for the prevention of disasters and the mitigation of its effects are undertaken by the departments of the Government at the district level as well as by the local authorities.
- Ensure that the guidelines for prevention of disasters, mitigation of its effects, preparedness and response measures as laid down by the National Authority and the State Authority are followed by all departments of the Government at the district level and the local authorities in the district.
- Give directions to different authorities at the district level and local authorities to take such other measures for the prevention or mitigation of disasters as may be necessary.
- Lay down guidelines for prevention of disaster management plans by the department of the Government at the districts level and local authorities in the district.
- Monitor the implementation of disaster management plans prepared by the Departments of the Government at the district level.
- Lay down guidelines to be followed by the Departments of the Government at the district level for purposes of integration of measures for prevention of disasters and mitigation in their development plans and projects and provide necessary technical assistance.
- Monitor the implementation of measures.
- Review the state of capabilities for responding to any disaster or threatening disaster situation in the district and give directions to the relevant departments or authorities at the district level for their upgradation as may be necessary.
- Review the preparedness measures and give directions to the concerned departments at the district level or other concerned authorities where necessary for bringing the preparedness measures to the levels required for responding effectively to any disaster threatening disaster situation.

- Organise and coordinate specialised training programmes for different levels of officers, employees and voluntary rescue workers in the district.
- Facilitate community training and awareness programmes for prevention of disaster or mitigation with the support of local authorities, governmental and nongovernmental organisations.
- Set up, maintain, review and upgrade the mechanism for early warnings and dissemination of proper information to public.
- Prepare, review and update district level response plan and guidelines.
- Coordinate response to any threatening disaster situation or disaster.
- Advise, assist and coordinate the activities of the Departments of the Government at the district level, statutory bodies and other governmental and nongovernmental organisations in the district engaged in the disaster management.
- Lay down guidelines for, or give direction to, the concerned Department of the Government at the district level or any other authorities within the local limits of the district to take measures to respond effectively to any threatened disaster situation or disaster.
- Coordinate with, and give guidelines to, local authorities in the district to ensure that pre-disaster and post-disaster management activities in the district are carried out promptly and effectively.
- Provide necessary technical assistance or give advise to the local authorities in the district for carrying out their functions.
- Review development plans prepared by the Departments of the Government at the district level, statutory authorities or local authorities with a view to make necessary provisions therein for prevention of disaster or mitigation.
- Examine the construction in any area in the district and if it is of the opinion that the standards for the prevention of disaster or mitigation laid down for such construction is not being or has not been followed, may direct the concerned authority to take such action as may be necessary to secure compliance of such standards.

- Identify buildings and places which could, in the event of any threatening disaster situation or disaster, be used as relief centers or camps and make arrangements for water supply and sanitation in such buildings or places.
- Establish stockpiles of relief and rescue materials or ensure preparedness to make such materials available at a short notice.
- Provide information to the State Authority relating to different aspects of disaster management.
- Encourage the involvement of non-governmental organisations and voluntary social-welfare institutions working at the grassroots level in the district for disaster management.
- Ensure communication systems are in order, and disaster management drills are carried out periodically.
- Perform such other functions as the State Government or State Authority may assign to it or as it deems necessary for disaster management in the District.

Role of Policy Makers in Disaster Risk Reduction

- Obtain commitment from public auth.
- Increase public awareness & public participation through education programmes.
- Fostering better understanding of causes & impact of disasters.
- Create legislative instruments.
- Comprehensive urban development strategies & land use plan.
- Legally mandate of building codes.
- Enforce laws effectively.
- Develop techno-financial regime.
    - Allotment of funds for disaster mitigation.
    - Development & establishment of early warning (EW) systems.

Bodies engaged in Animal Welfare in India

- AWBI (Animal Welfare Board of India)
- SPCA (Society for the Prevention of Cruelty to Animals)
- MOEF (Ministry of Environment and Forests)
- SAHVD (State Animal Husbandry & Veterinary Departments)
- AWO (Animal Welfare Organizations, NGOs)

❑❑❑

National Disaster Response Force (NDRF)

Raising of NDRF

- The disater profile of country compelled Government of India to formulate a disaster management plan.
- DM act was passed by parliament on 26th dec, 2005.
- NDMA constituted in the year 2005 to supervise the disaster management plan of the country.
- Chapter–VIII, sec-44(1) of DM act asked for creation of national disaster response force (NDRF).
- As per MHA order dated 19/01/2006, 08 battalions, 02 each from BSF, CRPF, ITBP and CISF were raised.
- On 14th feb, 2008, NDRF rules were notified by Govt. of India.
- Till sep. 2009, all these 08 battalions were under administrative & operational control of their respective DGS.
- As per MHA order dated 11 sep, 2009 command and control of NDRF to be exercised by DG NDRF.
- During 2010, 01 battalion each of BSF & CRPF were raised & converted into NDRF, thus, making total 10 battalions of NDRF.

NDRF Concept

- 10 NDRF battalions
- A specialist response force with:
  - High skill training;
  - State of the art equipments.
- A multi disciplinary, multi skilled and high tech force:

- For all types of disaster capable of insertion by Air, Sea & Land.

- All NDRF battalions to be equipped and trained for all natural disasters including NBC.
- Dedicated exclusively for disaster response
- NDRF battalions located to cover most vulnerable and high risk areas within 4-8 hrs in the country.
- Geared to take on Regional and International commitments on short notice.
- Ready and pro active availability of NDRF to States.
- Trained for heli-insertion in impact areas.
- Readiness of force to be on highest priority.
- NDRF battalions to facilitate all possible targets for state response forces.
- NDRF will be a dedicated force exclusively for disaster response and capacity building activities related to disaster management.

Vision

- As per cabinet note NDRF to be raised on the lines of NSG as a specialist stand alone rapid response force for manmade & natural disasters.
- Equipped with all state of the art equipments.
- Highly trained and skilled force to respond to any kind of disaster within the country & abroad.

Objectives of NDRF

- To work for community capacity building for disaster response.
- To trained the state disaster response forces.
- Deployment in case of impending disasters.
- Provide specialised response in case of disasters.

Role of NDRF

Pre-Disaster

- Will impart Disaster Management training to the local police, District Authorities, PRIs and other Stake holders.
- Will organize community awareness and preparedness programme during familarisation & recece of vulnerable areas in the A.O.R.

- Will impart training to the teachers and school children in the management of Natural Disasters.

During-Disaster

- Pro-active deployment at level–III disaster.
- Provide hi-tech-specialist response during search and rescue of victims.
- Provide immediate relief with available men and material in the affected areas.

Post-Disaster

- To assist States in recovery- maintenance of normalcy at the areas of disaster.

Composition of NDRF battalions

- Each battalion has 1149 personnel.
- Each battalion has 18 specialist teams of 44 members to handle natural & NBC disasters.
- Each team has Engineers, Paramedics, Technician, Electrician, Communication personnel & Dog squad.
- Organized, equipped and trained for all type of disasters.

Locations of NDRF in India

1. Bhatinda
2. Ghaziabad
3. Patna
4. Guwahati
5. Kolkata
6. Bhubaneshwar (Mundali)
7. Vadodara
8. Pune (Talegoan)
9. Vijayabara
10. Chennai (Arakkonam)

The location of NDRF centre at Ghaziabad, Kolkata, Pune (Talegoan) and Chennai (Arakkonam) having facilities of natural as well as manmade disaster management especially CBRN.

Functional Parameter

- Regular and intensive training.
- Regular familiarization of A.O.R.
- Carryout mock drills.
- Carryout joint exercise with various stakeholders.
- To carryout community awareness / preparedness programme in area of responsibility.
- To train states police organizations.

Best Practices

- On 12 Nov 1970, a major cyclone hitcoastal belt of Bangladesh at 223 km/hr, 500,000 people died.
- Due to cyclone preparedness plan the April 1991, cyclone with wind speed of 225 km/hr killed only 138,000 people even though the coastal population had doubled by that time.
- In May 1994, in a similar cyclone with wind speed 250 km/hr, only 127 people lost their lives.
- In May 1997, in a cyclone with wind speed of 200 km/hr only 111 people lost their lives.

Future Possibilities

- Mobilising stakeholders participation of Self Help Groups, Youth Groups, Panchayati Raj Institutions.
- Anticipatory Governance: Simulation Exercises and Scenario Analysis.
- Indigenous knowledge systems and coping practices.
- Living with risk: Community Based Disaster Risk Management.
- Inclusive, Participatory, Gender Sensitive, Child Friendly, Eco Friendly and Disabled friendly disaster management.
- Technology driven but people oriented.
- Knowledge management: Documentation and dissemination of good practices.
- Public Private Partnership.

Incident Command System

Incident: An incident is an occurrence, either caused by human or natural phenomena, that requires response actions to prevent or minimize loss of life, or damage to property and/or the environment.

Incident Command System (ICS)

- Is a standardized, on-scene, all-hazard incident management concept.
- Allows its users to adopt an integrated organizational structure to match the complexities and demands of single or multiple incidents without being hindered by jurisdictional boundaries.

ICS Purposes

Using management best practices, ICS helps to ensure:

- The safety of responders and others.
- The achievement of tactical objectives.
- The efficient use of resources.

History of ICS

Weaknesses in incident management were due to:

- Lack of accountability.
- Poor communication.
- Lack of a planning process.
- Overloaded incident commanders.
- No method to integrate interagency requirements.

The identification of these areas of management weakness resulted in the development of ICS.

ICS Benefits

- Meets the needs of incidents of any kind or size.
- Allows personnel from a variety of agencies to meld rapidly into a common management structure.
- Provides logistical and administrative support to operational staff.
- Is cost effective by avoiding duplication of efforts.

Basic Features of ICS

- Common terminology;
- Modular organization;
- Integrated communications;
- Management by objectives;
- Reliance on an Incident Action Plan (IAP);
- Chain of command and unity of command;
- Unified Command;
- Manageable span of control;
- Designated Incident Facilities.

Common Terminology

Using common terminology helps to define:

- Organizational functions.
- Incident facilities.
- Resource descriptions.
- Position titles.

Modular Organization

- Develops in a top-down, modular fashion.
- Is based on the size and complexity of the incident.
- Is based on the hazard environment created by the incident.

Management by Objectives

- ICS is managed by objectives.
- Objectives are communicated throughout the entire ICS organization through the incident planning process.

Overall Priorities

Incident objectives are established based on the following priorities:

- Life Saving;
- Incident Stabilization;
- Property Preservation.

Reliance on an Incident Action Plan

Every incident must have an Incident Action Plan (IAP) that:

- Specifies the incident objectives.
- States the activities to be completed.
- Covers a specified timeframe, called an operational period.
- May be oral or written except for hazardous materials incidents, which require a written IAP.

Elements of an Incident Action Plan

Every IAP must have four elements:

- What do we want to do?
- Who is responsible for doing it?
- How do we communicate with each other?
- What is the procedure if someone is injured?

Unity of Command

Under unity of command, personnel:

- Report to only one supervisor.
- Receive work assignments only from their supervisors.

Don't confuse unity of command with Unified Command!

Unified Command

- Enables all responsible agencies to manage an incident together by establishing a common set of incident objectives and strategies.
- Allows Incident Commanders to make joint decisions by establishing a single command structure.
- Maintains unity of command. Each employee only reports to one supervisor.

Manageable Span of Control

Span of control

- Pertains to the number of individuals or resources that one supervisor can manage effectively during an incident.
- Is key to effective and efficient incident management.

ICS Management: Span of Control

ICS span of control for any supervisor:

- Is between 3 and 7 subordinates.
- Optimally does not exceed 5 subordinates.

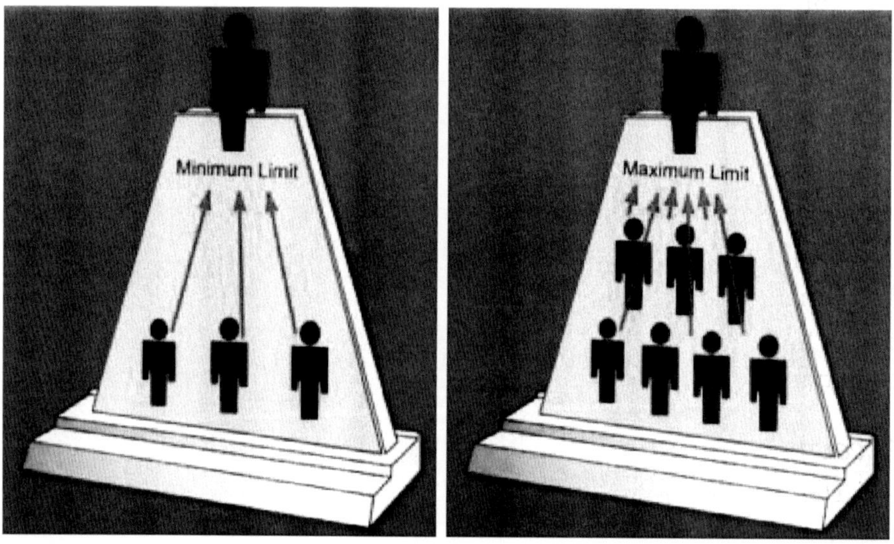

Predesignated Incident Facilities

Established by the Incident Commander based on the requirements and complexity of the incident.

Resource Management

Resource management includes processes for:

- Categorizing resources.
- Ordering resources.
- Dispatching resources.
- Tracking resources.
- Recovering resources.

It also includes processes for reimbursement for resources, as appropriate.

Integrated Communications

Incident communications are facilitated through:

- The development and use of a common communications plan.
- The interoperability of communication equipment, procedures, and systems.
- Before an incident, it is critical to develop an integrated voice and data communications system (equipment, systems, and protocols).

Transfer of Command

Transfer of command occurs when:

- A more qualified person assumes command.
- The incident situation changes over time, resulting in a legal requirement to change command.
- There is normal turnover of personnel on extended incidents.
- The incident response is concluded and responsibility is transferred to the home agency.

Accountability

The following principles must be adhered to:

- Check-In: All responders must report in to receive an assignment in accordance with the procedures established by the Incident Commander.
- Incident Action Plan: Response operations must be coordinated as outlined in the IAP.
- Unity of Command: Each individual will be assigned to only one supervisor.
- Span of Control: Supervisors must be able to adequately supervise and control their subordinates, as well as communicate with and manage all resources under their supervision.
- Resource Tracking: Supervisors must record and report resource status changes as they occur.

Incident Commander Role

The Incident Commander

- Provides overall leadership for incident response.
- Delegates authority to others.
- Takes general direction from agency administrator/official.

Incident Commander Responsibilities

The Incident Commander is specifically responsible for:

- Ensuring incident safety.
- Providing information services to internal and external stakeholders.
- Establishing and maintaining liaison with other agencies participating in the incident.
- Is responsible for all activities and functions until delegated and assigned to staff.
- Assesses need for staff.
- Establishes incident objectives.
- Directs staff to develop the Incident Action Plan.

Incident Commander

The Incident Commander performs all major ICS command and staff responsibilities unless the ICS functions are delegated and assigned.

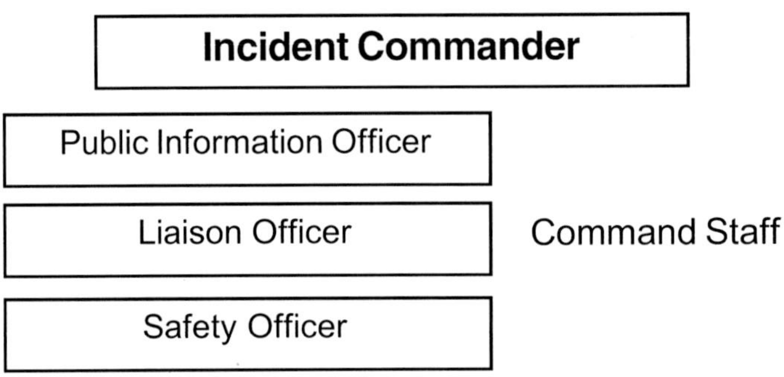

General Staff

Operations Section Chief	Planning Section Chief	Logistics Section Chief	Finance/Admin Section Chief

Command Staff

It may be necessary for the Incident Commander to designate a Command Staff:

- Provide information, liaison and safety services for the entire organization.
- Report directly to the Incident Commander.

Public Information Officer (PIO)

- Advises Incident Commander on information dissemination and media relations. Incident Commander approves information that the PIO releases.
- Obtains information from and provides information to Planning Section.
- Obtains information from and provides information to community and media.

Liaison Officer (LO)

- Assists Incident Commander by serving as point of contact for agency representatives who are helping to support the operation.
- Provides briefings to and answers questions from supporting agencies.

Safety Officer (SO)

- Advises Incident Commander on issues regarding incident safety.
- Works with Operations to ensure safety of field personnel.
- Ensures safety of all incident personnel.

General Staff

- General Staff in the ICS organizational structure includes operations section chief, planning section chief, logistics section chief and finance/admin section chief.

Operations Section Chief (OSC)

The Operations Section Chief:

- Develops and implements strategy and tactics to carry out the incident objectives.
- Organizes, assigns, and supervises the tactical field resources.
- Supervises air operations and those resources in a Staging Area.

Operations Section

- Directs and coordinates all incident tactical operations.
- Is typically one of the first organizations to be assigned to the incident.
- Expands from the bottom up.
- Have the most incident resources.
- May have Staging Areas and special organizations.

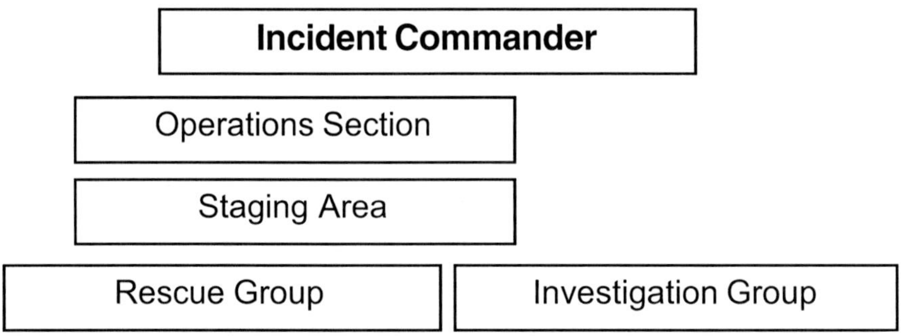

Planning Section Chief (PSC)

The Planning Section Chief:

- Gathers, analyzes, and disseminates information and intelligence.
- Manages the planning process.
- Compiles the Incident Action Plan.
- Manages Technical Specialists.

Planning Section

- Maintains resource status.
- Maintains and displays situation status.

- Prepares the Incident Action Plan.
- Develops alternative strategies.
- Provides documentation services.
- Prepares the Demobilization Plan.
- Provides a primary location for Technical Specialists assigned to an incident.

Logistics Section Chief

The Logistics Section Chief:

- Provides resources and services required to support incident activities.
- Develops portions of Incident Action Plan and forwards them to Planning Section.
- Contracts for and purchases goods and services needed at the incident.

Logistics Section

Responsible for:

- Communications.
- Medical support to incident personnel.
- Food for incident personnel.
- Supplies.
- Facilities.
- Ground support.

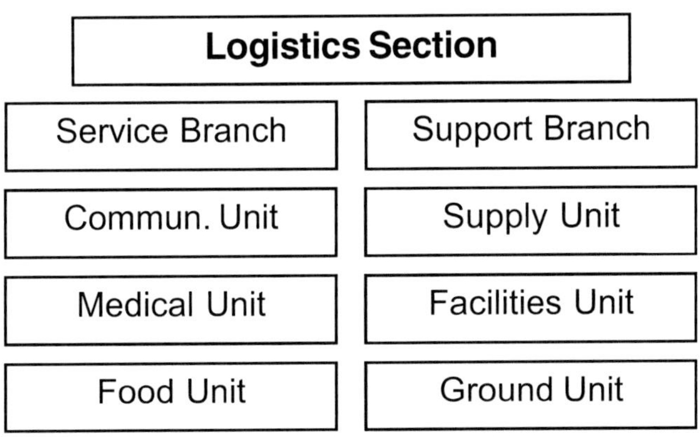

Finance/Administration Section Chief

The Finance/Admin Section Chief:

- Is responsible for financial and cost analysis.
- Oversees contract negotiations.
- Tracks personnel and equipment time.
- Processes claims for accidents and injuries.
- Works with Logistics to ensure resources are procured.

Finance/Administration Section

- Contract negotiation and monitoring
- Time keeping
- Cost analysis
- Compensation for injury or damage to property

Finance/Administration Section

Time Unit	Compensation/ Claims Unit
Procurement Unit	Cost Unit

Predesignated Incident Facilities

Incident facilities are:

- Established by the Incident Commander depending on the requirements and complexity of the incident or event.
- Activated only when needed. Some incidents may require facilities not included on the standard list.

Incident Facilities: Incident Command Post

Incident Command Post (ICP)

- Is the location from which the Incident Commander oversees all incident operations.
- May change locations during the event.

- May be located in a vehicle, trailer, tent, or within a building.
- Should be positioned outside of the present and potential hazard zone but close enough to the incident to maintain command.
- Every incident must have some form of an Incident Command Post.

Incident Facilities: Staging Area

Staging Areas

- Are temporary locations at an incident where personnel and equipment are kept while waiting for tactical assignments. The resources in the Staging Area are always in available status. There may be more than one Staging Area at an incident.
- Should be located close enough to the incident for a timely response, but far enough away to be out of the immediate impact zone.
- May be collocated with the ICP, Bases, Camps, Helibases, or Helispots.

Incident Facilities: Base

Base

- Is the location from which primary logistics and administrative functions are coordinated and administered. There is only one Base per incident, and it is designated by the incident name.
- May be collocated with the Incident Command Post.
- Is established and managed by the Logistics Section. The resources in the Base are always out of service.

Incident Facilities: Camps

Camps

- Are where resources may be kept to support incident operations if a Base is not accessible to all resources. Multiple Camps may be used, but not all incidents will have Camps.
- Are temporary locations within the general incident area that are equipped and staffed to provide food, water, sleeping areas, and sanitary services.
- Are designated by geographic location or number.

Incident Facilities: Helibase/Helispots

- A Helibase is the location from which helicopter-centered air operations are conducted. Helibases are generally used on a more long-term basis and include such services as fueling and maintenance.
- Helispots are more temporary locations at the incident, where helicopters can safely land and take off. Multiple Helispots may be used.

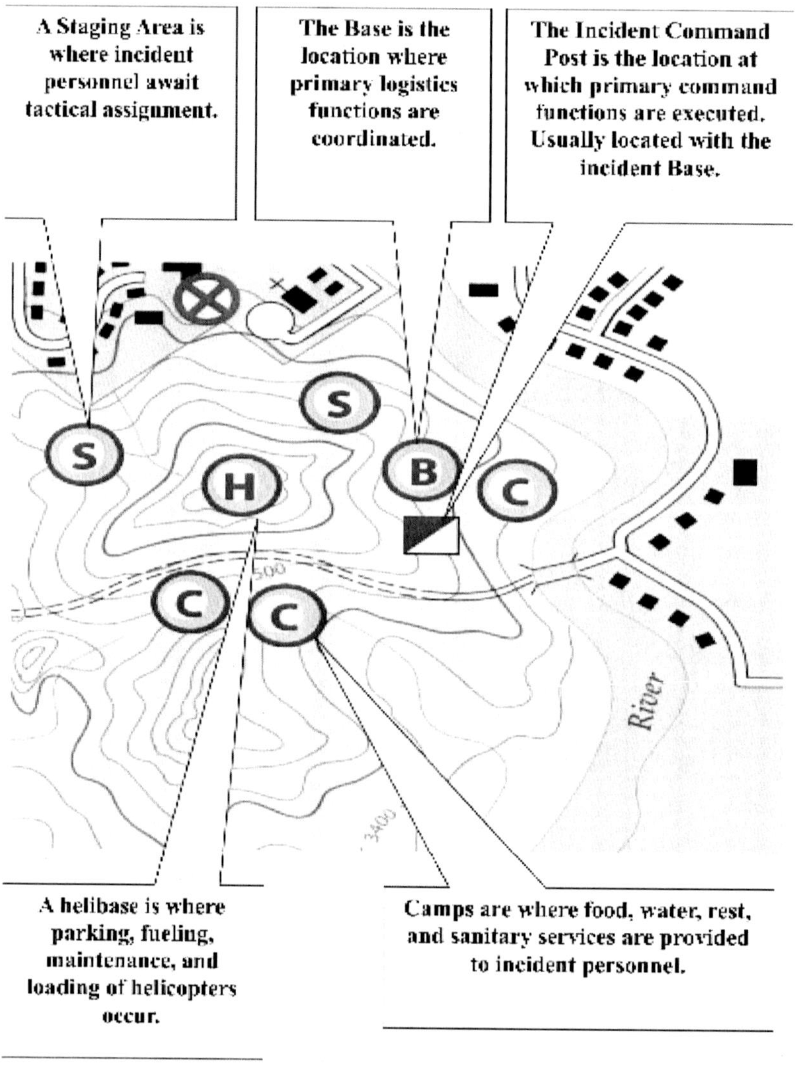

❑❑❑

Basic Life Support and Cardio Pulmonary Resuscitation

Two third of the victims of heart attack died out of hospital, due to lack of knowledge and modern cardiac facilities. Most of the death occurs within two hours of onset of symptoms of heart attack. Respiration and heart beats are our life line; in absence of either our body has a tendency to cease other causing brain cell die. CPR (Cardio Pulmonary Resuscitation) is a vital link in the chain of survival. There are four links of chain of survival. The chances of survival are more if they come together. These links are as follows:

- *Early Access* : Recognize the signs and symptoms of cardiac and respiratory emergencies early and notify emergency medical services as early as possible.
- *Early CPR* : Performing effective CPR as early as possible after onset of illness. Starting CPR early increases greatly the chances for survival of the victim.
- *Early Defibrillation* : If trained on and equipped with automated external defibrillator **(AED),** use it as soon as possible. This link is the most likely to improve survival rates.
- *Early Advanced Care* : It is important that advanced medical care or advance cardiac life support (ACLS) be available rapidly for a positive outcome.

These interventions is not limited to the victims of heart disease, but also useful in drowning, trauma, electrocution, suffocation, air way obstruction, allergic reaction etc., may be saved by prompt intervention There are certain risk factors which causes heart problem. Such factors can be divided into three groups. The greater the prevalence of risk factors, the greater the likelihood of heart disease or other blood vessel disease.

Factors that cannot be changed are :

1. Family history
2. Sex
3. Ethnic background
4. Age

Risk Factors that can be changed

Smoking, High blood pressure, High Cholesterol, Physical activity

Contributing factors: Obesity Diabetes, Diabetes, Excessive stress

The Cardiovascular System

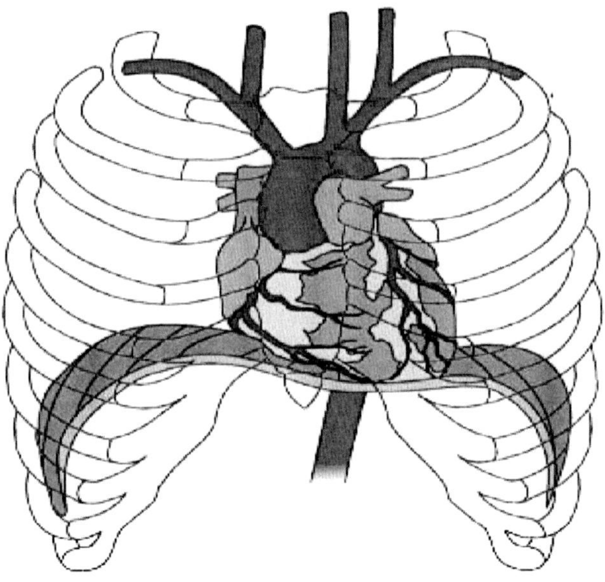

Function of Lungs and Heart

- The cardiovascular system consists of the HEART, ARTERIES, CAPILLARIES & VEINS. The heart is a muscular organ , approx the size of a fist, and is located in the thoracic cavity behind the sternum and between the lungs. The coronary arteries are special arteries the supply blood to the heart muscles themselves.
- The function of the Heart is to Pump Blood.
- The Left side receives oxygenated blood from the lungs & pumps it to the body through the arteries. The right side receives. From

the veins, the blood that has circulated through the body and pumps it to the lungs to be oxygenated once again.

- A system of one way valve keeps the blood flowing in right direction and prevents it from flowing backward.

Respiratory system : It is made up of four components:

- An airway (upper and lower).
- A neuromuscular system (includes the respiratory centre in the brain, respiratory muscles and the nerves that connect the two).
- *Alveoli* – tiny air sacs surrounded by capillaries. The alveoli are surrounded by the capillaries. The brain sends nerve signals to muscles in the thorax and diaphragm, causing to breathe. With each inhalation, air is carried through the airways to the alveoli in the lungs, where oxygen & carbon dioxide are exchanged.
- Arteries, capillaries and veins.

Breathing

Adequate breathing is characterized by :

- Chest and abdomen rise and fall with each breath.
- Air can be heard and felt exiting the mouth or nose.
- Ease of breathing (effortlessness).
- Adequate rate.

Inadequate breathing is characterized by

- Inadequate rise and fall of the chest.
- Noisy breathing: Bubbles , rales. Stridor , whistling etc.
- Increased respiratory effort.
- Cyanosis.
- Inadequate rate.
- Altered mental status.

Sign of Absent Breathing

- No chest or abdominal movement.
- Air cannot be heard or felt exiting the mouth or nose.

Cyanosis

- A bluish coloration of the skin and mucous membranes caused by the lack of oxygen in the blood and tissues is called Cyanosis. This condition can be the result of the patient breathing in an environment poor in oxygen, suffering from illness or Respiratory injury, or A/W obstruction. It can also be noticed on the lips, Ears and nostrils or nail beds. In patients with dark pigmentation, it is necessary to inspect the nostrils, palms and nail beds and the mouth and tongues.

Clinical & biological death

- The Respiratory and Circulatory system are interdependent - If either one stop; the other will do the same in a very short time. The Brain is the 1st organ to suffer the effects of a lack of oxygen. Shortly after oxygen supply is cut off, brain cells began to die, causing irreversible damage.
- Clinical death occurs when a patient is in respiratory arrest (not breathing) or in cardiac arrest (heart not beating). The patient has a period of 4 to 6 minutes to be resuscitated (revive) without brain damage. It can be reversed.
- Biological death occurs at the moment when the brain cells begin to die. It cannot be reversed.
- *Exception*: Cold water drowning. There have been cases of persons resuscitated one hour or more after cold water drowning. In the cases victims should receive prolonged resuscitative efforts. In a cold environment, a person should not be considered dead until the victim's body is warmed.

Signs of Certain Death

- *Lividity* : The pooling of blood in the lower areas of the body.
- *Rigor mortis* : Stiffening of the body and limbs that occurs after death, usually within 4 - 10 hours.
- *Decomposition* : A decomposing body always produces a fetid odour. The rate of decomposition depends on a number of factors, primarily ambient temperature.
- *Other signs* : Mortal wounds such as decapitation, dismemberment, incineration, severe crushing injuries etc. Only a medical doctor can pronounce a person officially dead.

Techniques for opening the air way

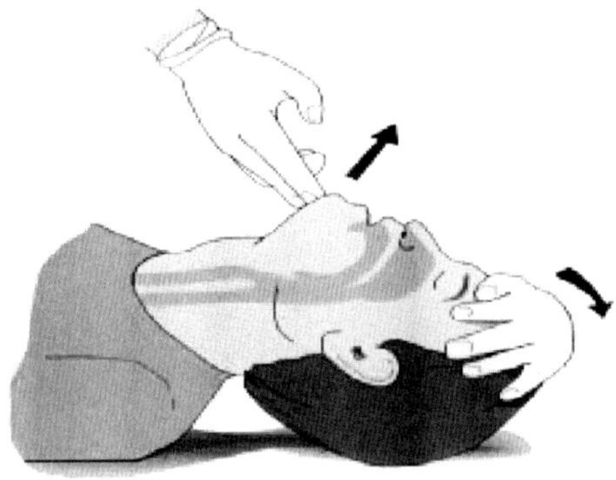

Steps of Head Tilt Chin Lift

- Position the patient lying face up
- Kneel by patient?s shoulders towards the head.
- Place one hand on the forehead and place the fingertips of your other hand under the bony part of the patient's jaw.
- Lift up on the chin, supporting the jaw, and at the same time, tilt the head back as far as possible. For infants and children: place in the sniffing position don't over extend.
- Always keep the patient's mouth slightly open. Use your thumb to hold down patient?s lower lip.
- Never dig into the soft tissue under the patient's chin.
- Once the airway is open, check breathing by LLF. If patient is not breathing start artificial ventilation. If unable to ventilate, assume the air way is obstructed.

Artificial Ventilation by opening A/W by Head -Tilt Chin – Lift method

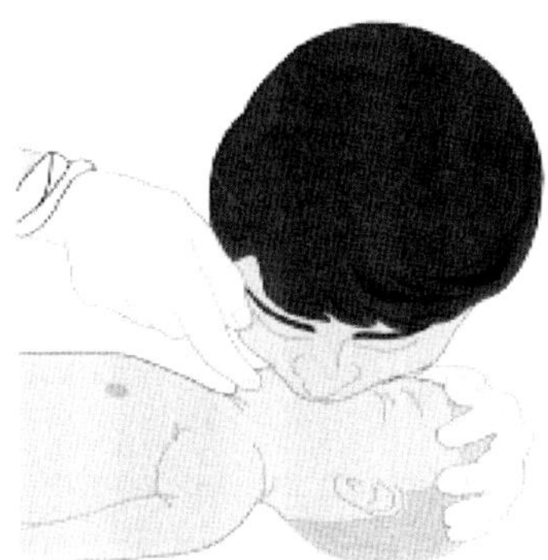

Air way opening by jaw thrust method

- It is the only maneuver recommended on an unconscious patient with suspected Head, Neck or Spinal injury.

Step of Jaw Thrust

- Position the patient laying face up.
- Kneel above the patient's head.
- Place your elbows next to the patient's head on the surface where he is laying.
- Place both hands on either side of the patient's head.
- Grasp the angle of the patient's jaw on both sides, for an infant or child use two to three fingers.
- Use a lifting motion to move the jaw forward (up) with both hands.
- Keep the patient's mouth slightly open by using your thumbs if needed.

Artificial Ventilation

- Once the patient has an open airway you can provide artificial ventilation for a patient breathing inadequately or not at all.
- Natural air contains 21% oxygen and the body only utilizes about 5%. Therefore exhaled air contains 16% oxygen. This exhaled air can resuscitate a person who is not breathing, until a high-concentration oxygen source is available. There are Three Techniques of artificial breathing.
- *Mouth to Mask* : This method uses a pocket face mask with a one way valve to form a seal around the mouth & Nose of patient. This method is preferred as it eliminates direct contact with the patient and prevents exposure. Place the mask around the mouth and nose of the patient. The narrower top portion of the mask

should be seated on the bridge of the nose. The broader portion should fit the chin. Seal the mask by placing heel and thumb of each hand along the border of the mask and placing firmly to provide a tight seal around the edges of the mask. Open A/W by using proper method. Give the breath at proper rate and duration observing chest rise and fall.

- *Mouth to barrier device* : There are two barrier device : Mask & Shield. Position the barrier device around the patient's mouth and nose, proving an adequate seal. Open A/W by using proper method. Give the breath at proper rate and duration observing chest rise and fall.

- *Mouth to Mouth* : Mouth to mouth ventilation is very risky in the field as it may cause infection to the rescuer. It is a personal decision whether to perform or not. Open A/W by using proper method. Gently pinch the nose of the patient with your thumb and index finger to prevent air from escaping. Take a deep breath and seal your lips around the patient's mouth, proving an adequate seal. In case of infant and child cover both the nose and mouth with your mouth. Give the breath at proper rate and depth.

Breathing rates and duration

- Adults : 10-12 breaths per minute lasting 1.5 to 2 sec
- Children and Infants : 20 breaths per min lasting 1-1.5 sec.
- New borns : 40 breaths per min lasting 1-1.5 sec.
  - Look proper chest rise. With infants and new borns, use puff from mouth to avoid over-ventilate.

Hazards to rescuers during artificial ventilation

- *Disease* : Air/blood borne. Mask, gloves, and eye protection may be worn. Use BVM or pocket mask.

- *Chemicals* :Exposure from a contaminated patient. Patient should be decontaminated first.

- *Vomitus* : One way valve on a pocket mask or BVM should be used.

Risk to Victims during artificial ventilation Gastric Distention

- This problem can occur during rescue breathing, which can force some air into the patients stomach, causing the stomach to became inflated, or distended. This can result in two serious problems:
  - Reduced lungs volume-The lungs is upwardly displaced by the diaphragm.
  - Vomiting - Explosive expulsion of fluids or partially digested food from the stomach into the throat resulting into airway obstruction, aspiration of vomit into lungs, possible causing lung damage and/or a lethal form of pneumonia.

Prevention of Gastric Distention : Avoid or minimize gastric distention by positioning the patient's head properly and by avoiding giving ventilations that are too forceful or too quick. Volume should be limited to that which causes the chest to rise adequately.

- Foreign Body Airway Obstruction.
- *Causes of Airway Obstruction* : There are two types of A/W obstruction based on location of foreign body i.e. upper & Lower A/W obstruction.
  - An upper A/W obstruction is anything that blocks the back of the mouth or throat or the nasal passage.
  - A lower A/W obstruction is caused by breathing in a foreign body or by severe spasm of the bronchial passage such as asthma.

Besides above based on effects of obstruction, it is divided into two groups.

- *Partial Air way Obstruction* : An object that caught in the throat that partially block breathing. Followings are the signs of this obstruction.
  - Inadequate or poor air exchange.
  - Weak ineffective cough.
  - High pitch noise when inhaling.
  - Increased respiratory difficulty and may clutch at the throat
  - Cyanosis (Bluish discoloration of the skin and mucous membrane).

- *Complete Air Way Obstruction* : A patient with complete foreign body airway obstruction will not be able to speak, breath or cough. May clutch his neck with thumb and finger this gesture is known as the Universal sign of Choking. Movement of air will be absent.

Air way Obstruction is caused by

- *Tongue* : Tongue falls back blocking the throat. Commonly in unconscious.
- *Foreign Body* : Objects like food, toys, vomits etc.
- *Acute epiglottises* : When patient tries to force breathing. Also caused by allergies.
- *Tissue damage* : Due to penetrating injury to neck, crushing trauma to the face, inhalation of hot air, ingestion of chemicals.
- *Illness* : Respiratory infection and certain chronic conditions (such as asthma).
- Most common A/W obstruction in a responsive patient is food and in the unresponsive patient it is tongue.
- *Managing FBAO in Adults and Children* : The method recommended for relieving FBAO with poor air exchange or complete obstruction is the abdominal thrust.
- Each thrust intended to relieve obstruct.
- Perform several thrust until FB is expelled.
- It may damage the internal organs.
- To minimize damage always keep your thrust between xiphoid process and naval.

Abdominal Thrust-Responsive Adult or child (Patient sitting or standing)

- Perform an abdominal thrust : Grasp the fist the other hand and press the fist into the patient's abdomen with a quick upward thrust.Repeat thrust until the object is expelled from the airway or **patient becomes unconscious.** Each new thrust should be a separate and distinct movement.
- Managing FBAO in adult and children.

Abdominal Thrust Unresponsive adult or child (Lying down)

- Position the patient face up (supine).
- Attempt to ventilate, If unsuccessful reposition the patient's head and try again. If still unsuccessful, go to the next step.
- Get in position. Kneel astride the victim's thighs and place the heel of one hand against the patient's abdomen, midline, slightly above the naval and below the XP. Place the 2nd hand directly on top of the first.
- Perform up to five thrusts. Press the abdomen with a quick upward thrusts.

- Perform a finger sweep. In children, use this technique only if you are able to see the foreign object. Use the tongue jaw lift to open the patient's mouth, Insert the index finger of the other hand along the inside of the cheek into the throat, using a hooking action to dislodge the FB and lift it out.
- Repeat steps 2 through 5 until patient's airway is opened.

Chest Thrust- Pregnant or Obese Responsive Adult (patient standing or sitting)

- Chest thrust are to be used only with patients in late stages of pregnancy or with markedly obese, when abdominal thrust can not be applied effectively.
- Get in position. Stand behind the patient with your arms directly under the patient's armpit and encircle the the patient's chest.
  - Position your hands Place the thumb –side of your fist along the patient's sternum, avoiding XP and margins of the rib cage.
  - Perform a chest thrust. Grab your fist with the other hand and perform backward thrust until and until the object is expelled from the airway or the patient becomes unconscious. Repeat this step until the A/W is opened.

Chest Thrust- Pregnant or Obese Unresponsive Adult (patient Lying Down)

- Position the patient: Place the patient on his/her back and kneel close to the patient's side.
- Attempt to ventilate. If unsuccessful, reposition the patient's head and try again. If still unsuccessful continue with following.
- Get in Position. Place the heel of your hand on the lower half of the sternum of the patient. Place your other hand on the top of the first. Do not press on the XP.

Chest Thrust- Pregnant or Obese Responsive Adult (patient standing or sitting

- Deliver up to five chest thrust. Each thrust must be distinct downward motion.

- Perform finger sweep. Use the tongue jaw lift to open the patient's mouth. Insert the index finger of the other hand along the inside of the cheek into the throat, using a hooking action to dislodge the FB and lift it out.
- Repeat steps 2 through 5 until patient's airway is opened.

Managing FBAO in Infants

Always suspect a FBAO in infants who demonstrate a sudden onset of respiratory distress associated with gagging or wheezing. Most common causes are ; Toys or Other small objects. As mentioned earlier, airway obstruction may also be caused by infection suspect this condition if infant has a fever with congestion, hoarseness or drooling. Do not attempt to relieve this form of obstruction and transport the patient immediately. Verify complete airway obstruction: Serious breathing difficulty, ineffective cough, no strong cry. Position the infant: Straddle the infant face down over one of your forearms, head lower than the body. Support the infant's head by holdinf the jaw with your hand. Deliver 5 back blows . Use the heel of your hand between the shoulder blades. If FO is not expelled , position the infant face-up on your arms. Head lower than the body. Deliver 5 chest thrust. Position your middle and ring fingers in the middle of the infant's sternum, just below the imaginary line between the infant's nipples. Use a quick downward motion. Repeat steps 2 to 5 until effective, or until the infant becomes unconscious. Establish unresponsiveness.

- Open A/W and ventilate.
- If still obstructed. Reposition the infant's head and ventilate again.
- Give up to 5 back blows and 5 chest thrusts.
- Perform a tongue – jaw lift. If you can see the foreign object, Use a finger sweep to remove it.
- Repeat steps 2-4 until effective.

Chapter 7

Psycho-Social Support in Disasters

Psycho-social Support in Disasters

India : High Risk

- Floods, droughts, cyclones, earthquakes found it every year.
- Communal riots, conflicts, refugees, epidemics and other disasters compound the country's chronic troubles.
- Other types of manmade hazards add to the tirade of problems.

India Disaster Profile

8% Cyclones

12 % floods

28% droughts

56% Earthquakes

Landslides

Tsunami

India is a theatre of disasters !!!

India - Disaster Profile

- Industrial disasters-Bhopal Gas Tragedy
- Epidemic outbreaks -Plague
- Fires -Dabwali, Uphar Cinema
- Air accidents-Patna
- Railway accidents-Konkan
- Fire tragedy
- Venus Circus Fire Tragedy
- Kumbakonam School Fire Tragedy
- Fishing village fire tragedy in Rameshwaram

Disaster : Disaster is a severe disruption of ecological and psycho social which greatly exceeds the coping capacity of affected community (WHO, 1992).

Two types of disaster

Natural disaster: which are due to nature's fury.

Human made disaster: Because of human involvement in the cause of disaster.

- Human intension cause higher problems.
- Human error comparatively causes fewer problems.

Response to Disaster

Disaster : Magnitude, occurrence, type

Community : Level of preparedness, past experiences, Leadership, Influences Community reactions

Survivor : Age, gender, exposure, Personality, coping skills, Social supports, losses, Health status

Response : Adaptive, Maladaptive

Disaster leads to

- Displacement, loss, death and destruction.
- Disruption of person place, relationship.
- Imbalance of psychological well being.
- The disturbance in nurturing environment.

Natural and Human Made Disasters : In natural disaster the epicenter is one whereas in human made disaster the epicenters are many.

Normal Reactions for Abnormal Situations

- Guilt
- Fear
- Shock
- Intrusive memories
- Grief
- Vigilance
- Numbness
- Despair

Psycho-Social Impact: It is two types

1. Short term

- Giddiness / instability
- Panic attacks
- Generalised anxiety

- ❐ Somatic symptoms
- ❐ Emotional problems
  - irritability,
  - apprehension,
  - numbness
- ❐ Increased alcohol use

Effects of Disasters

Physical Impact	Economical Impact	Social Impact	Psychological Impact
• Decreased Sleep and Appetite • Fatigability • Susceptible to infectious diseases • Death • Nausea, vomiting • Diarrhea • High temperature • Infections • Respiratory problems • Abortion • cholera, chicken	• Loss of Livelihood/ property • Damage to Livestock • Lack of adequate employment opportunity • Increased financial burden • Loss of breadwinner • Family Disharmony • Migration • Decrease in standard of living	• Problems in inter personal relationship • Loss of parents, single parents , widow/widower • Communication problems • Collapse of social structure • Loss of social status • Education, recreation and social activities • Reduction in productivity • Environmental and hygiene problems • Spread of diseases • Property loss /Lack	• Anxiety and fear • Horror and disgust • Stunned and numb • Anger and scapegoting • Paranoia • Loss of trust • Demoralisation • Hopelessness • Helplessness • Guilt • Somatic complaints

2. Long term

- ❐ Pathological grief
- ❐ PTSD (Post Traumatic Stress Disorder)
- ❐ Depression
- ❐ Panic & anxiety disorders
- ❐ Substance abuse disorder

Post Traumatic Stress Disorder (PTSD)

- PTSD is a condition that develops after exposure to a terrifying event or deal in which grave physical harm occurred or was threatened.
- The event must have involved actual or threatened death or serious injury and caused an emotional reaction involving intense fear, hopelessness, or horror.
- People with PTSD have three kinds of experiences for weeks or months after the event is over and the individual is in a safe environment.
- *Re-experiencing/intrusion*: PTSD describes a syndrome in which trauma survivor is unable to get the event in question out of his/her mind.
- *Avoidance/numbing*: Survivors develop a set of avoidance behaviors that are designed to restrict their exposure to reminders of the event.
- *Hyperarousal*: They have trouble sleeping, are irritable, have impaired concentration, are hypervigilant, and have increased startle responses.

Impact of Disaster

- Psycho-social impact is always greater than the physical/medical.
- Can affect all spheres of life.
- Has a long range ripple effect.

Social Issues

- Displacement of the individual.
- Tracing and reunion issues.
- Details about the family members.
- Refusal to accept the loss/injury.
- Confusion about the nature of the injury & treatment
- Hopelessness and helplessness.

Lifestyle Changes

- Worse change in lifestyles are associated subsequent to a disaster.
- Tobacco use.

- Alcohol consumption.
- Family violence.
- Sleeping hours.
- Wife battering.
- Inability to work.
- Lesser nutrition.
- Multiple tasks.

Tenants of DM & PSS

- Higher impact of event.
- Higher the distress.
- Lesser functionality.
- Loss of coping capacity.
- Social fabric disrupted.
- Eroded social support.
- Need for PS care, support and rehabilitation.

Psycho-social Care Principles

From the international disaster literature the following broad principles of care has been identified:

- No one who experiences the event or witnesses the event is untouched by the event.
- Disaster results in two types of trauma, namely the individual and the collective trauma.
- Most people pull together and function during and after a disaster but their effectiveness is diminished.
- Disaster stress and grief reactions are normal responses to an abnormal situation.
- Support systems are crucial for recovery.

Psychosocial experiences

Psycho-social Care (PSC) - Spectrum of Care

- Reconstruction initiatives need to integrate psycho social care as an essential part of the overall care and rehabilitation.
- Those who receive a spectrum of care would have less emotional reactions, better functionality and quality of life.

Psycho-social Support Stages

A –Psychosocial first aid

B –Psycho social care

C –Psycho social rehabilitation

Lifetime –rebuilding of the family and community.

Steps of Psycho-social First Aid

- Meet immediate needs
- Listen, Listen, Listen
- Accept feelings
- Assist with next step
- Refer and follow up

PSC TECHNIQUES FOR ADULTS

TECHNIQUES OF PSYCHO-SOCIAL CARE FOR DISASTER SURVIVORS

EVERYONE CAN FACILITATE RECOVERY

PSC Mediums for Children

Psycho-social Care for Children

- Singing
- Stories
- Games
- Painting
- Drawing
- Modelling
- Play
- Drama
- Puppetary

Women-Survivors of Disaster

- Lack of privacy in camps to discuss intimate details.
- Further exploitation to vulnerable women.
- Tend to talk about physical problems than the psychological stress.
- Increased distress.
- Vulnerability of women in disasters.

Human Animal Bond

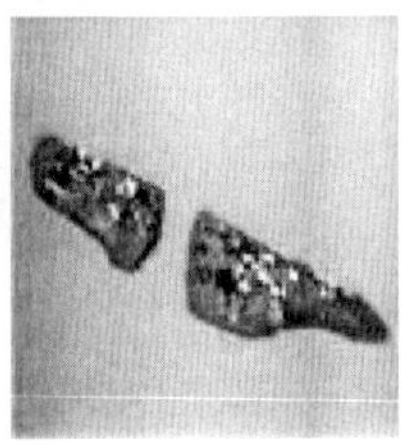

Help in Disaster

- The human connection to animals influences compliance with individual and community evacuation plans.
- Search and rescue teams with canine units confront physical and emotional demands that affect both handler and animal.
- The culling of animal populations on a large scale affects every member of rural community.
- Livestock farmers and their families endure enormous emotional losses, and veterinarians and government officials who must implement these programs suffer as well.

Veterinarian Role is Very Important

- The animal-human bond may affect human beings 'psychological and behavioral responses in disasters.
- It is critical to integrate an awareness of mental health needs in veterinarian disaster response in order to develop effective evacuation plans and manage the emotional impact of controlling livestock epidemics that require the slaughter of healthy animals.
- The grief experienced over losing a beloved pet or herds maintained for generations is often unrecognized or minimized in our society.
- Veterinary emergency response teams must be sensitive to these experiences and understand their role in sustaining the psychological well-being of members of rural farm communities as well as themselves.
- There is a need to further examine the relationship of animals and humans in actual disaster response.
- Understanding the intricate relationship between humans and animals is an important component of a comprehensive public health approach to disaster response and a critical element in promoting the resilience of individuals and communities.

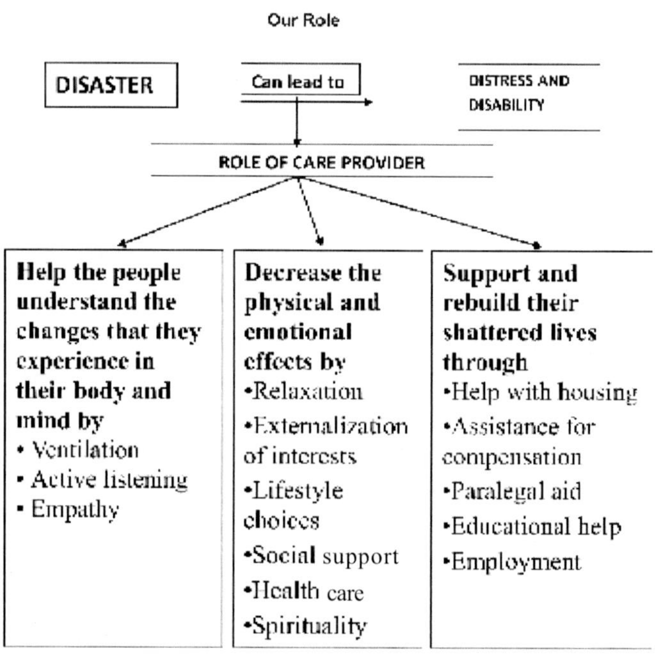

All the above three lead to:

↓

ADJUSTMENT AND MASTERY OVER EMOTIONAL DISTRESS

Responder Stress and Well-Being: Helping the Helpers

Psycho-social impact on the care givers

- Not a normal work.
- Suddenness and enormity and demanding.
- Multiple challenges.
- Dealing with the loss, painful experiences is challenge.
- Their own loss and exposure also will increase the stress.
- First responders -emergency management personnel.
- Their role is time bound.
- Pressure ....
- Change in the place of work.
- Continues demand from the people.
- Inability to get serious results.
- Pressure to fulfill the community as well organizational mandate.
- They are not immune.
- Survivors need over takes care givers.

Self care for care givers

- Understanding and recognizing stress.
- No quick fix solutions.
- Stress management is health promotion and preventive.
- Predominantly in group form, typically standardized.
- Prime goal is to teach skills, be educational; benefits may take time to show.

Visibility and invisibility the understanding developed

- Physical disability is more visible.
- Helps in reducing the level of actual and perceived stress and in preventing adverse psychological and social consequences among disaster affected people.

STRESS CYCLE

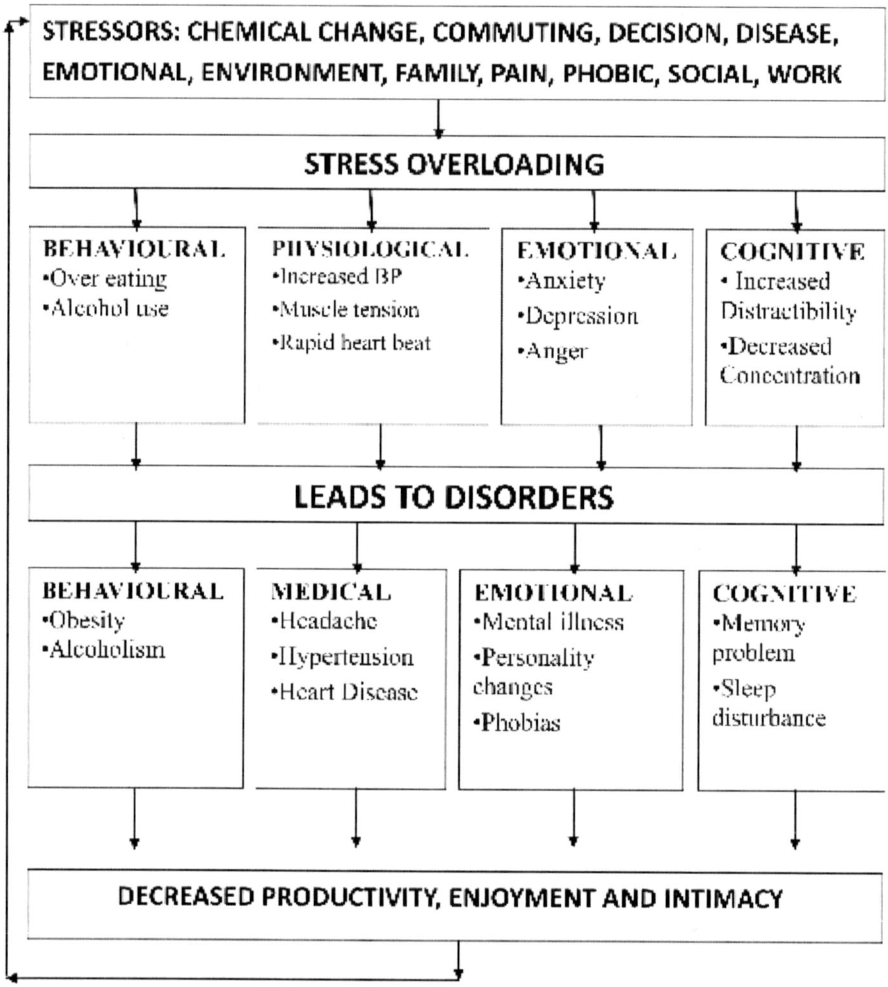

MODEL OF STRESS

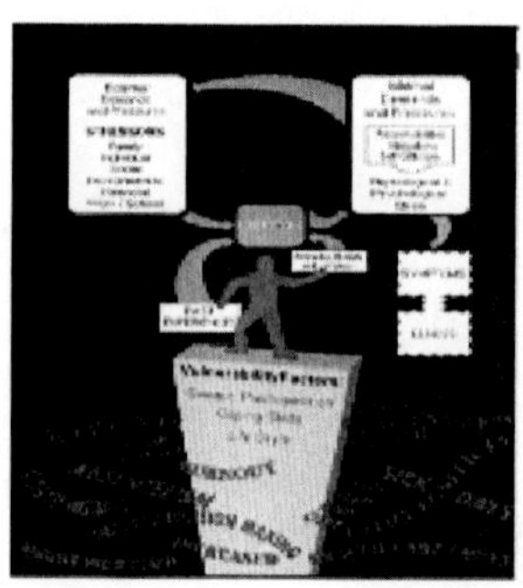

- Trained psychosocial care giver needs to pay close attention to identify these needs.

Psycho-social care experiences

- Moving from clinical to public health and psycho social care models.
- From Psychiatry to Normalcy paradigm.
- Involving and Empowering community resources/personnel.
- Increasingly higher number of stake holders and field players.
- Emergence of evidence based research.

Mental health resources

- Very Scarce -37 Mental Hospitals, 160 General Hospital Psychiatric Units.
- Present status of available mental health personnel will be 3500 psychiatrist, 500 clinical psychologist, 400 psychiatric social workers and 1000 psychiatric nurses.
- 8 -10 % of our population needs psychiatric help...
- In disaster the needs multiply.

Professionals per 100, 000 Population	USA	UK	AUSTRALIA	INDIA
Number of Psychiatrist	13.7	11	14	0.2
Number of Psychologist	31.1	9	5	0.03
Number of Social Workers	35.3	58	5	0.03
Number of Psychiatric Nurses	6.5	104	53	0.05

Source: Mental Health Atlas 2005

Need for institutionalization

- Providing psycho-social support in disasters is NOT institutionalized.
- Centre and states governments have been deploying few mental health professionals in disasters.
- Significant development during the last decade has been extensive involvement of international agencies, NGOs–Gujarat earthquake, Tsunami and Kashmir earthquake.
- There is no uniformity in the Psycho-social support delivery of services.

Paradigm Shift

- Shift in thinking and focus from reactive disaster response and relief centric approach to a **HOLISTIC APPROACH** towards Preparedness, Prevention and Mitigation for an effective and prompt Response.
- The shift stresses on an '**all hazard**' disaster management multi sectoral/multidisciplinary approach – all aspects of DM cycle.

❑❑❑

Climate Change and Disaster Risk Reduction

Climate Change : It is a broader term that refers to long-term changes in climate, including average temperature and precipitation.

Global Warming : It is the increase of the Earth's average surface temperature due to a build-up of greenhouse gases in the atmosphere.

History of Earth's Climate

- Earth formed ~4.6 billion years ago.
- Originally very hot.
- Sun's energy output only 70% of present.
- Liquid water present ~4.3 billion years.
- Life appeared ~3.8 billion years ago.
- Photosynthesis began 3.5-2.5 billion years ago.
- Produced oxygen and removed carbon dioxide and methane (greenhouse gases).
- Earth went through periods of cooling ("Snowball Earth") and warming.
- Earth began cycles of glacial and interglacial periods ~3 million years ago.

Green House Gases (GHG) Emission

Agricultural Production Increase at the cost of....

- CO_2 emission 1.4-1.7 GtC (10-12% of total anthropogenic GHG emission).
- N_2O emission 0.76 GtC (58% of total anthropogenic GHG emission). Nitrogenous fertilizer account for 38%.
- CH_4 emission 0.90 GtC (47% of total anthropogenic GHG emission).

  Livestock enteric fermentation and manures 38%.

  Rice fields 11%.

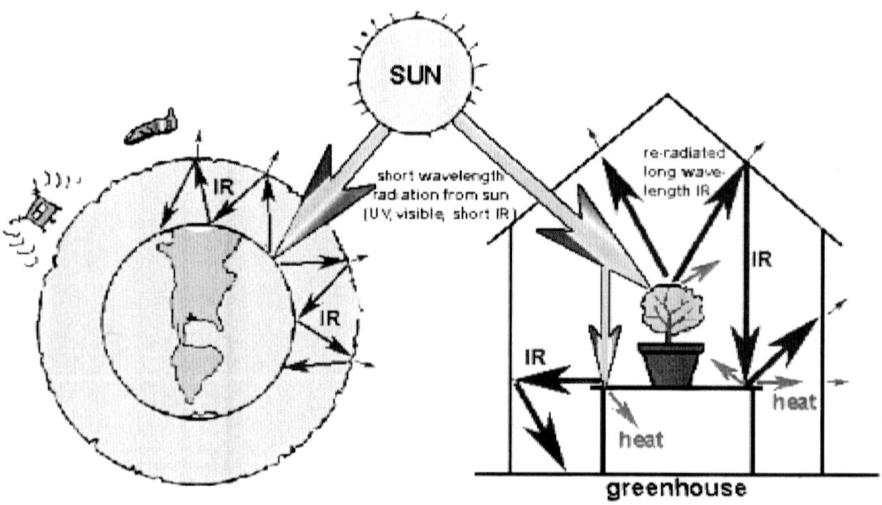

- Carbon dioxide stays in the atmosphere for approximately 100 years, methane lasts about 12 years. Other greenhouse gases last even longer.
- As these gases continue to raise surface temperatures, they trigger the release of even greater quantities of carbon dioxide and methane that are currently trapped in frozen Arctic permafrost and tundra soils, further increasing temperatures.

Achievements since Green Revolution

- Global cropland increased by 27%.
- Population increased by 111%.

- Food production increased by 164%.
- Crop yields increased by 139%.

Impact of Climate Change

- *Production*: Reduced crop yield, particularly in south Asia.
- *Soil*: Drier, reduced productivity.
- *Irrigation*: Increased demand, reduced supply.
- *Pests*: Increased ranges and populations.
- *Livestock*: Increased diseases and heat stress.
- *Fishery*: Affected abundance and spawning.
- *Economic impact*: Reduced agricultural output.

Water Availability - a Global Threat

Blue-tendency to increase

Red-tendency to decrease

Water availability in Asian Countries:

Countries	1950	2000	2010	2025	2050
	Per capita water availability (m^3 $annum^{-1}$)				
Bangladesh	56411	16744	14335 (75)	12086	10593
China	5047	2210	2068 (59)	1927	1976
India	5831	2000	1717 (71)	1457	1292
Indonesia	31809	12325	10881 (66)	9609	8781
Malaysia	74632	19593	16336 (78)	13503	11497
Nepal	21623	6958	5695 (74)	4470	3467
Pakistan	11844	3159	2533 (79)	1900	1396
Philippines	15390	4158	3450 (78)	2754	2210
Thailand	8946	2871	2627 (71)	2465	2440
Vietnam	12553	4780	4223 (66)	3684	3367

Source: Gardner-Outlaw and Engelman (1997)

- In Dryland areas 1300 m^3 yr^{-1}.
- Below the minimum average requirement (2000 m^3 yr^{-1}).

Soil Health- Degradation

- Declining organic matter.
- Reduced microbial activity.
- Reduced soil aggregation.
- Soil compaction and poor drainage.

Climate Change

Carbon dioxide emissions: Projected CO_2 emissions under different population and technology assumptions, 1990 to 2100 (billions of metric tons of carbon)

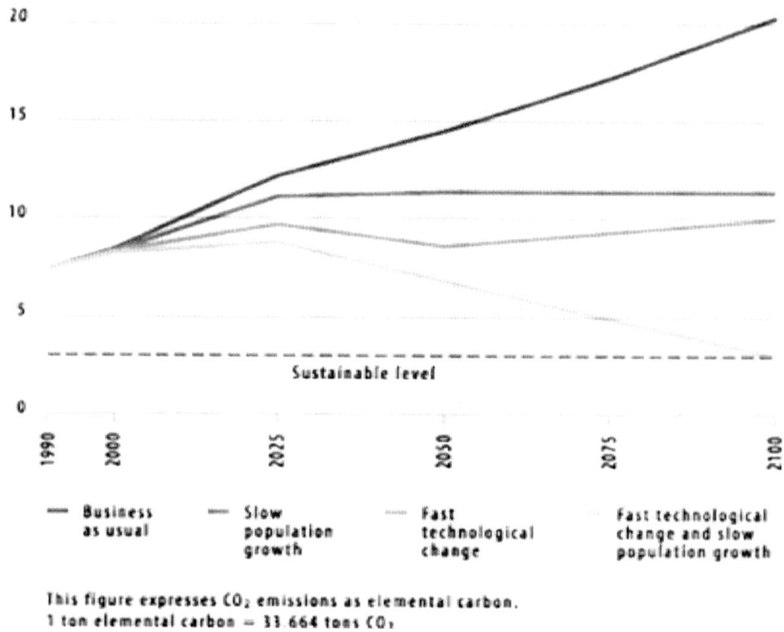

Climate Change Scenarios for India

Year	Season	Increase in Temperature (°C)		Change in Rainfall (%)	
		Lowest	Highest	Lowest	Highest
2020s	Rabi	1.08	1.54	-1.95	4.36
	Kharif	0.87	1.12	1.81	5.10
2050s	Rabi	2.54	3.18	-9.22	3.82
	Kharif	1.81	2.37	7.18	10.52
2080s	Rabi	4.14	6.31	-24.83	-4.50
	Kharif	2.91	4.62	10.10	15.18

Mitigation strategies

Sink and Sequester Enhancements

- Carbon sequestration in soils
- Changes in land use and management: higher biomass.

Emission Reductions

- CO_2 reduction by reducing fuel usage, tillage operations.
- N_2O reduction by better management of nitrogenous fertilizer.
- CH_4 reduction from rice fields and livestock.

Avoid Emission

- Prevent deforestation.
- Substitution of biomass based energy for fossil fuel energy.

Adaptation strategies

Increase Resilience

- Diversity of cropping and varieties.
- Diversity of production.
- Timely field operations and higher flexibility.
- CA based practices for drought, rain, heat, Conservation Agriculture = sustainable profit.
- Soil cover/residue management.
- In-situ moisture and nutrient.
- Reduced water losses.
- Soil cover help to counteract heat, cold, frost, erosion.
- Improve SOC.
- Help in water recharge, infiltration and quality.
- Better input efficiency.
- Pest and disease management.
- Weed management.

Crop diversification helps in climate change adaptation and risk mitigation

Sustainable Rural Development for Disaster Mitigation

The disaster management planning at village level and utility of management tools in disaster mitigation planning for sustainable rural development:

Leadership

- For the success of any mission a leader is must.

Traits and style

- Drive and determination, Self confidence.
- Leadership motivation and knowledge.
- Courage.
- Flexibility.
- Honesty and integrity.

Developing Volunteers for Disaster Mitigation

- Training for volunteers.
- Awareness for people.

Search and Rescue Team

- One of the most important components during any sort of disaster.

Incident Command System

- Information, safety and liaison unit.
- Operational, planning, logistic and Finance Unit.

System to be adopted

Village → Block → District → State → National level → International

Factors affecting Search and Rescue

- Hazardous tools and equipment.
- Vandalism and theft.
- Extreme fatigue and stress.
- Adverse weather condition.
- Unfamiliar worksite.
- Extreme situation like fire, dust, smoke.

Communication network for Disaster

Radio spectrum

- Very high frequencies (V.L.F.) 30-3000MHz-T.V., F.M., Radar Air Navigation.
- Short wave is the most suitable instrument for communication

Mobile, Land phone etc.

Village level activity

- Plantation.
- Water storage/tank.
- Road linkages.
- Communication setup.
- Involvement of villagers.
- Earthquake/Fire resistant houses.

Urban level activity

- RCC structure.
- Fire proof house.
- Water storage/pipe line near slum.
- Avoid mass gathering.
- Help in accident case.
- Awareness for Motivation.

Arrangement for mass awareness through:

- Involvement of all community.
- Government and public interaction.
- Training and seminar/workshop for Medicos/Vets/Police/ Block development officials/panchayet/Sarpanch/Public representative.

Role of NGOs and professional societies in sustainable rural development

Non-Governmental Organisations (NGOs) and Private Voluntary Organisations (PVOs) are by definition non-profit making organisations. Some are religious, others secular, some are national, others of foreign origin. However, they all have one thing in common - their aim - alleviation of human sufferings and development. It must be borne in mind that NGOs/PVOs do not mean either large private foundations or private firms, which are in search of profit, though they too are clearly both private and non-governmental organisations. NGOs and PVOs mobilise the enthusiasm and commitment of volunteers for disaster mitigation on one hand and for developmental activities on the other.

Natural disasters such as famine, earthquake, drought and floods and man-made disasters such as riots and fires move them into action. They also swing into action through educational efforts to draw the attention of the public to the crying needs of the society such as arresting population explosion, disaster preparedness, environmental protection etc. These organisations are, however, best known for their relief and rescue activities. As against this, it is observed that the Government finds it very difficult to manage and mitigate disasters; as it tends to take upon itself many responsibilities that could have been better left to agencies like NGOs and PVOs, who could handle the job. In addition to NGOs, which can be treated as an organised sector, there is an unorganised sector of the local level individual experts that could also help in the cause of disaster management and sustainable development.

Any Government, however well-equipped it might be, cannot all alone handle the problems of disaster management and economic development. Its efforts have to be supplemented by social, educational, professional and voluntary organisations, which possess tremendous potential and therefore can help in large scale peoples involvement. Exploitation of full potential of these organisations has, however, remained untapped so far, mainly on account of lack of coordination and inspiration. Role of civil society and professional societies have gained paramount importance in the present decade.

History of voluntary efforts in India

Origin of voluntary efforts in India, as elsewhere, can be traced back to social service with its antecedents in charity and, therefore, has close links with social reform and missionary activity aimed at spreading enlightenment and alleviating sufferings. In the earlier era, education, medical care and concern for the physically and socially disabled people were among the preferred fields of service.

The activities and anxieties of voluntary agencies turned to relief measures during two successive drought years in India in the mid-sixties. Flood, famine and cyclone relief programmes had been undertaken even earlier too, but nothing was as large and sustained as efforts mounted to fight the famine in Bihar and elsewhere. This was also a land mark in the scale of foreign funding, entry of foreign NGOs and other forms of international support. Contributions in the form of food grain, milk-powder, clothes, medical supplies, drilling rigs, etc., started pouring in the large quantities. Eventually, the Government decided to channelize these donations through an official "Freedom from Hunger" campaign office in the Ministry of Food and Agriculture. This later on became the "People's Action for Development in India (PADI) an official agency with non official representation, for allocating the rupee counterpart of foreign donations in kind - food grain and other food supplies, to Indian field agencies on the basis of approved programme.

During the British period such activities were discussed and sponsored through clubs, as a legacy of which even today we see Lions Club and Rotary International pursuing such activities. These clubs, where the government officers and well to do people in the society, met in the evening, provided a forum for discussion for the ills of the society and what could be done to remove them. Several schemes were taken up as a follow up on these discussions both by the government departments and by the voluntary organisations through contribution.

Side by side with these clubs, came up, professional societies like the Institution of Engineers, Indian Medical Association, etc., whose aims are limited to bringing together the people belonging to the particular profession and to arrange discussions about developments related to their profession.

Classifications of NGO's

NGO's can, be broadly classified as under:

- Educational institutions.
- Voluntary agencies.
- Multi-disciplinary societies.

- Professional societies.
- Micro-specialised societies.

Educational Institutions: The country today has a largest network of educational institutions like schools, colleges, universities, research centres, etc., functioning in the non-governmental sector side by side with the government institutions. These institutions however have so far not played any significant role at the local level planning except for some institutions associated for conducting services for Integrated Rural Development Programme (IRDP). They have remained unconcerned and uninvolved with the social problems related to disaster management and sustainable development. Some of the social work that many schools take up are also mainly in the field of campaign for literacy, hygiene, etc. This large population of educated youth could be used in educating the public on disaster mitigation, availability of natural resources, developmental programmes, environmental impact assessment and need for conservation as a part of their school curriculum either under special social service or socially useful productive works (SUPW). Participation of over a million student volunteers apart from NCC, National Service Scheme, Nehru Yuvak Kendras, Bharat Scouts and Guides can make success of rescue, relief and rehabilitation in disaster besides creating proper public awareness.

Voluntary Agencies: There are many voluntary agencies in the country today engaged in the task of relief work and rural development. The earliest among these were the Sewagram set up by Mahatma Gandhi and Shantiniketan set up by Rabinder Nath Tagore. These agencies due to their better understanding of local needs, problems and resources, can evolve new programmes, strategies and approaches through local people in fulfilling many social needs without much expenditure. These agencies are of different sizes. Some of the larger agencies which cover from 40 to 100 villages through one or more programmes are Sewa Mandir, Kishore Bharti, Gandhi Smark Nidhi, Agrindees, Smanvaya Ashram and Tamulpur Anchalic Gram Dan Sangh, Xavier Institute of Social Services. The programme of these agencies is however not uniform. General education and welfare programmes tend to cover more villages and population rather than other programmes. The associations set up on the Gandhian pattern such as Gandhi Vichar Prishad, Gram Bharti, Gandhi Samarak Nidhi have common objective including socio-economic and cultural development, undertaking constructive programmes with which

Gandhiji was associated, developing individual, family and community and creating awareness among people about social problems. The major field of their activity, however, could be grouped into disaster mitigation, education, irrigation & agriculture, cottage & village industries and health.

Multi-disciplinary Societies: These societies are those which have among their members people from all walks of life rather than a particular profession who have direct or indirect interest in disaster mitigation, developmental projects and environmental management. Some of such notable multi-disciplinary societies in India are:

1. Assam Science Society.
2. Marathi Vighyan Parishad.
3. Kerala Sahitya Parishad.
4. Indian Association for Environment Management.
5. Indian Institute of Public Administration.
6. Indian Environment Congress Association.
7. Indian Society of Training and Development.
8. Industrial & Chambers Federations like ASSOCHAM, CII, FICCI, PHDCCI.
9. Institute of Directors.
10. Society for Environmental Conservation.
11. World Environment Foundation.

These societies are well organised and have been successfully carrying out seminars, workshops, exhibitions, etc. for dissemination and propagation of knowledge and have many eminent people amongst their members. They are capable of taking up for study to advise on various problems and projects for disaster mitigation. These societies not only create suitable mass awareness but also deliberate upon to evolve consensus on future strategy in planning and tender suitable advice to the government at crucial times.

Professional Societies: Professional societies and institutions are those whose aims & objectives relate to a specific profession and their activities connected to the development in the concerned discipline. Some of the prominent Indian institutions/societies are:

1. Computer Society of India.
2. Confederation of Tourism Professionals of India.
3. Engineering Council of India.
4. Indian Medical Association.
5. Indian National Science Academy.
6. Institution of Engineers (India).
7. Indian Society for Technical Education.
8. Institute of Cost Works Accountants.
9. Institute of Chartered Accountants.
10. Indian Economists Association.
11. Indian Law Association.
12. Indian National Academy of Engineering.
13. Indian Council of Arbitration.
14. Indian Society of Valuers.
15. National Academy of Agricultural Sciences (NAAS).

The membership of these societies is limited mainly to the related professions. These societies are highly technical and have eminent scientists, engineers, doctors, economists, lawyers, chartered accountants, tourism professionals etc. as their members. Their opinion carries a lot of weight and often influences the policy decisions of the government. They however should gradually broaden their objectives with a view to take up activities involving general welfare of the society, particularly for disaster mitigation and sustainable economic development.

Micro-specialised Societies: The emphasis and need of micro-specialisation in various fields of engineering, science, agriculture, medicine and surgery, town planning, architecture, public administration, etc., have given birth to many professional societies. Their membership and activities are limited to a particular functional area of that profession. Such societies have many prominent experts as their members who strive hard to keep them abreast of the state of art in various challenging topics of that specialisation.

The societies in this group deliberate and publish papers covering research and developmental activities belonging to that micro profession. They can render considerable help in promoting awareness and advice to government on specific projects and problems of their concern.

Scientific knowledge and development NGOs

In this industrial age, the major concern for the relationship between human, natural resources and disasters mitigation has been the exponential growth of human population; whereas resources have been limited. Most of the NGOs should work through scientific information particularly the power of knowledge in terms of disasters and environment management. It is ridiculous to continue to believe in myths even; when they are clearly contradicted by scientific facts, and believing in two contradictory ideas. It is important to realise that most of the myths are created in the absence of precise knowledge of the matter. For better results, we need to enlarge our database which is evident in computers age. We can accordingly, manage disasters in better ways and resolve the mysteries of nature as well.

The only way to interact with nature is to enlarge our understanding of environment natural resources and development projects. NGOs should also understand relationship and functioning realising the limitations of human capabilities. However, every change in the environment should not be termed negative. The process of change is an integral part of the functioning of environment and natural resources management particularly due to very high consumption patterns in developed countries while population explosion in developing nations. The level of technological development influences the extent of human intervention influencing the magnitude of disasters and ecological status.

Earlier, NGOs were engaged in relief and charity, often under a religious/ missionary zeal feeding the hungry, supplying basic health care and education, focussing on the weak and vulnerable. Now, the focus shifted to development to evolve packages of technology, skills and market connections to facilitate better living and poverty alleviation. Soon, thereafter the range of activities of NGOs assumed a new complexity like advocacy for policy change, organising people for assertion of basic rights and the location of activity moved from local to the national and then to the global. NGOs have been praised for being in the forefront of micro credit provision for the poor, for pushing more inclusive and participatory planning and evaluation in raising public awareness.

The role of NGOs have been underlined in every major international forum eg., from Rio (environment), Vienna (human rights), Beijing (gender), Kyoto (climate change), Tehran (habitat), Copenhagen (social development) which still remains to be studied, particularly in their advocacy of 'abstract' international standards without adequate concern for time and context. It is a fact that most of the NGOs active in these domains are insufficiently rooted in the national or local context. To be genuine civic actors, it is crucial that the NGOs be both locally rooted and responsive. It is worth analysing why networking and lobbying for changes in a WTO forum (from Seattle to Cancun) comes more easily than lobbying for right concerns in their own country.

Another challenge is in forging relationships with other, independent entities not unlike coalition building. This is very difficult since few are willing to cede authority and control or even merge their objectives with others in service of a larger aim. Both globally and nationally; this appears like handing over to the better connected and endowed NGOs. The capacity building to develop new skills and learning for new roles in a changing context is also relevant. Finally there is the issue of legitimacy and accountability. Once the gap between precept and practice becomes glaring, it may well be that public ire may turn against NGOs. They must gear up to face this challenge.

Role for professional societies

It is a fact that very few people associated with the media reporting and environmental activism in India have proper understanding of the complexities of causes of disasters, environmental process and techniques of infrastructural development projects. The result is that they tend to overstress only one or two aspects of the problem.

Quite often the 'dose response function' of different disasters and environmental impacts are not understood, when looked at the processes in isolation. The causes and effect of natural disasters, environmental process and planning of sustainable developmental projects are quiet intricate and multi-disciplinary subjects.

In the recent decade, it is being noticed that a fear syndrome has been created by media and self acclaimed environmentalists against the developmental projects in general and hydropower development projects in particular through unjustified manifold exaggeration of the so called environmental impacts of such schemes. While we should ensure that environmental considerations are not neglected in the developmental process; nevertheless creating obstructions by exaggerating the minor environmental impacts side by side conveniently forgetting their tremendous benefits would be disastrous. Let us not forget that 40% of our people still live below the poverty line.

Mankind is at the centre of environment and if minimum needs of the society are taken care of, environment gets protected and sustainable development follows. Also as society's capacity to carry poverty is limited, economic development which reduces poverty becomes essential so that carrying capacity of natural resources remains within limits.

In the above difficult situation, all the professional societies would need to strengthen their forte for stimulating and supporting .the discussion on the real burning issues to serve the national interests in tune with the changing scenario. It is also essential that subjects of disaster management, environmental concerns and the process of planning and operation of various types of projects are rightly taught at different levels of education from an early stage as well as to the experts of different disciplines.

It is frequently felt that persons with true multi-disciplinary inclination are very few in our developmental and research organisations; most of the experts tend to develop a feeling that only they are able to diagnose the problems better. In the process, cooperative working arrangements and team spirit become causalities. Multi-disciplinary teamwork, though advocated easily, is indeed a difficult task to accomplish, specially when the individual's training is in the narrow disciplinary field of specialisation.

The success of multi-disciplinary team efforts relies heavily on effective team leadership. The team leaders should be able to sensitise and create enthusiasm for understanding and accommodating the needs of micro specialised disciplines, motivate team members, arbitrate differences, develop a basic understanding of all the disciplines involved and channelize the group efforts towards achieving the common goals.

Effective leadership also depends on the art of building good inter-personal relations, encouraging synergism and promoting group dynamics. Since structured university education hardly provides for exposure and development of skills in these areas, it becomes essential that the professional, multidisciplinary and specialised societies and voluntary agencies play their dominant roles in continuing education, quality up-gradation, disaster management and sustainable development through the following activities :

- Organising workshops, symposia, conventions etc. at national, regional and local levels.
- Publishing of Literatures in the form of Monographs, Papers, Conference Proceeding, Journals, etc.
- Conducting training programmes.
- Facility of libraries, reading rooms and resources centres.
- Distant learning facilities.
- Distribution of popular publicity material.
- Public awareness through mass media reporting.
- Technical and industrial exhibitions.
- Popular lectures from eminent experts.
- Regular classes for informal education and enhancement of qualifications.
- Award of Diplomas and Certificates after prescribed Examinations.
- Honouring and recognition of individual's achievements.
- Research and standardization.
- Organising functions like World Natural Hazards Mitigation Day, World Environment Day, World Water Day, World Tele-communication Day, World Health Day, Engineers Day, Water Resources Day, Teachers Day, etc., to focus discussions and activities on the pinpointed themes.
- Publishing rejoinders pointing out the scientific facts against unbalanced newspaper articles.
- Advising the governmental agencies through memorandums about the needs and perspective planning for disaster management and sustainable development.
- Participating in the subject committees formed at various levels to advise the government.

Unorganised effort

Apart from the organised efforts through NGOs and PVOs, many towns and villages have retired persons with technical, scientific, medical and other expertise, who may be willing to offer their services for preparing plans or schemes for the disaster management and sustainable developmental projects. These professionals possess a wealth of knowledge and information about their specialisation. Unfortunately, their special knowledge and expertise are hardly ever utilised. They are hardly consulted by the Panchayats or the government agencies.

On one hand there are complaints of the lack or inadequacy of expertise, on the other hand such large available resource remains untapped. There is a need of making a formal mechanism for associating locally available talent by drafting such persons to the advisory committees at the village, town, district and state levels. Such an association can help in obtaining data and information, carrying surveys, preparing schemes, formulating plans as a whole in the areas of preparedness, mitigation, monitoring, evaluation and management of natural disasters management.

Peoples Participation

Let us appreciate that disaster management and ecological degradation cannot be taken care of by a few experts and government agencies. Every individual has some role to play for sustainable development in a country like India. Proper orientation of people's mind is a pre-requisite in this task. The common man's mind is preoccupied with the problems of hunger and drinking water. He is too busy fighting with the vagaries of floods during monsoon and droughts during dry season.

Three legged system for sustainable development should be based on bringing awareness among people, continuous monitoring and interdisciplinary research and implementation of such findings. This would require close cooperation between government agencies and non-governmental organisations. A net work of environment awareness programme should include collection of information, analysis, evaluation and documentation followed by dissemination of information through lecturers, seminars etc. Hence, people's active involvement and suitable changes in their habits are of paramount importance for sustainability of economic developmental projects.

Commitment to Sustainability

Sustainable development should be regarded as a fundamental component of social responsibility, sound business practice and natural resource management. Sustainable development is development that meets the needs of the present without compromising the ability of future generations to meet their own needs.

Sustainable development requires the integration of three components - economic development, social development and environmental protection - as interdependent, mutually reinforcing pillars. Eradicating poverty, changing unsustainable patterns of production and consumption, and protecting and managing the natural resource base underpinning economic and social development are overarching objectives of, and essential requirements for sustainable development. Sustainable development is the collective responsibility of government, business, civil society, consumers and individuals. It is committed to working cooperatively with these sectors in achieving sustainable outcomes. We should care for the values of eco-efficiency and a precautionary approach to environmental management.

Eco-efficiency

We should support the opportunities to create more value with less environmental impact through the concept of eco-efficiency. Eco-efficiency is founded on the idea that becoming more efficient makes good business sense and is concerned with three broad objectives:

- Reducing the consumption of resources.
- Reducing the impact on nature.
- Increasing product or service value.

Precautionary Approach

A precautionary approach is one of the underlying values guiding efforts to achieve more sustainable outcomes for new and existing developments. In the application of this approach, public and private decisions should be guided by:

- Evaluation to avoid, wherever practicable, serious or irreversible damage to the environment;
- Consideration of the need for electricity and a reliable water supply to alleviate poverty and enhance living standards; and
- An assessment of the risks associated with various options.

Precautionary approach should be applied at the national and/or regional policy level. Decision makers should consider global issues, such as global warming, acid rain, and loss of biodiversity, when establishing national and/or regional policies for energy, water and land use. These issues should be addressed and dealt with at the national and/or regional policy level prior to the assessment of specific projects.

Conclusion

It is a challenge before us where the services of thousands of young men and women - highly qualified, with professional knowledge and commitment - who, presently, do-not fit in the existing governmental structure, can be utilised for the cause of disaster management and sustainable development. Professional societies and voluntary organisations can play a vital role in the preparedness, mitigation and management of natural disasters and optimum development of natural resources with due care of environmental management.

They may provide constructive criticism for timely correction and should side by side provide complimentary role by creating balanced scientific public awareness, informal educational and public participation. The modalities of the involvement and limitations of NGOs, educational, professional and voluntary organisations need to be worked out so as to streamline their useful participation in preparedness, mitigation and management of disasters.

We must remember that India has to support more than 1/6th of the world's population and 1/6th world's cattle, on just 1/50th of global land; with meager 1/25th of the world's water resources. In India, 80% of all diseases and over 1/3rd of deaths are caused due to contaminated water. World Bank has estimated that water pollution accounts for about 60% of the major annual environmental costs in India.

Safe water supply and environmental sanitation are vital for protecting environment, improving health and alleviating poverty Social tensions, political instability and street fights are already being experienced in India, on account of fast deteriorating availability of water in adequate quantity and of acceptable quality. Non-development of water storage and large hydropower projects is not a viable or available option; due to the very high temporal variations in river flows in Indian monsoon climate.

Environment is either science or engineering, which cannot be so well understood by novelists, journalists and self- styled activists. Such persons repeatedly use media to create fantasies since they have excellent command on language, media relationship and have nothing else to do except whole hearted full time tirade against large river valley projects, on many shifting grounds. Efforts of few non-professional activists in blocking large number of infrastructural projects should be curbed while replicating excellent examples of careful sustainable models like Delhi Metros, large number of river valley projects and planned cities like Chandigarh. It is also essential that water and energy sensitive planning is undertaken for cities and new urban centre.

We should promote greater consideration of environmental, social and economic aspects in the sustainability assessment of new infrastructural projects and the management and operation of existing schemes. Thorough sustainability assessments should ensure that detrimental social and environmental impacts are avoided, mitigated or compensated and positive outcomes are maximized. Of necessity, the principles are generic since each particular development project will have its own unique set of circumstances influenced by scale, geographic location, social, legal and political constraints.

□□□

Communication : An Effective Tool in Disaster Management

Communication

- It is the mutual interchange of ideas by any effective means.
- Agricultural Communication is defined as a "Planned transfer of farm technologies from the research system to farmers system through extra system and media with an objective of desirable changes in respect of higher productivity, profitability and prosperity and also get feedback from the clients".

Communication skills

- Communication skills is the ability to use language (receptive) and express (expressive information).
- Effective communication skills are a critical elements in your career and personal lives.

Communication goals

- To change behavior
- To get action
- To persuade
- To get and give information
- To ensure understanding

Process of Communication (Five Steps)

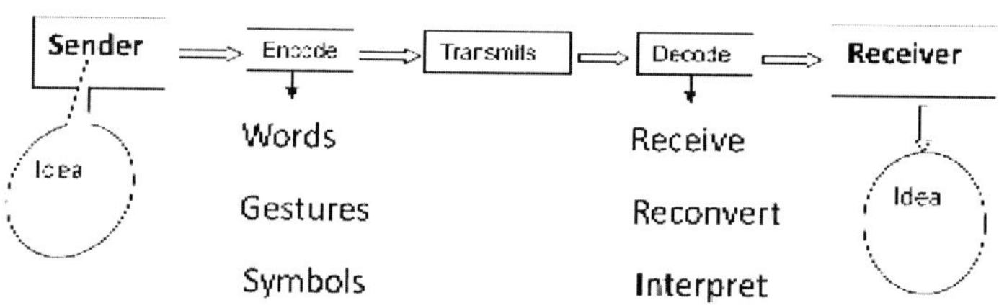

Factors Affecting Communication

- Communicator
- Message
- Channel
- Treatment of the Message
- Target Group/ Audience
- Effect / Audience Response

Communication is a series of experience of

- Hearing
- Smell
- Seeing
- Taste
- Touch

Most common ways to communicate

- Speaking
- Writing
- Visual image
- Body language

Barriers to communication

- Semnatic barriers
- Emotional or psychological barriers
- Organizational barriers
- Personal barriers

Semnatic barriers

- Symbols with different meaning
- Badly expressed message
- Faulty translation
- Unclarified assumption
- Specialist's language

Emotional or psychological barriers

- Premature evolution
- In attention
- Loss of transmission and poor retension
- Undue reliance on the written word
- Distrust of communication
- Failure to communicate

Organizational barriers

- Organizational policy
- Organizational rules and regulations
- Status relation
- Complexity in organization

Personal barriers

(a) Barriers in superior

- Attitude of superior
- Fear of challenge of authority
- Lack of time
- Lack of awareness

(b) Barriers in subordinates

- Unwillingness to communicate
- Lack of proper incentive

Communication involves three components

- Verbal message : the words we choose are brief, succinet, organized, free of jargon, do not create resistance in the listener.
- Paraverbal message : how we say the words such as tone, pitch and pacing of our voices.
- Nonverbal message : our body language such as facial expression, posture & gestures.

  These three components are used to send clear, concise messageds, receive and correctly understand messages sent to us.

Models of Communication

- Aristotle's Model

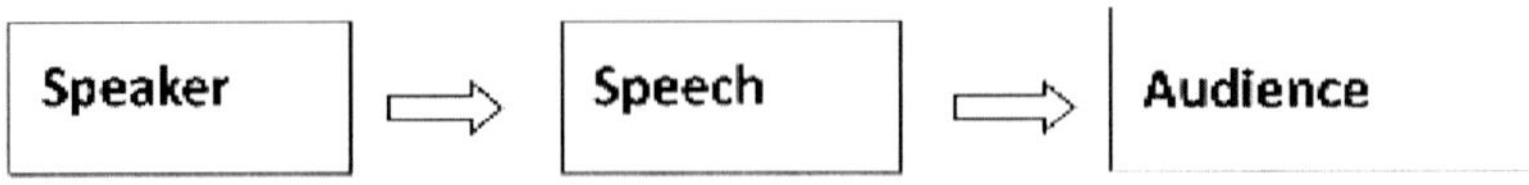

- Berlo's Model

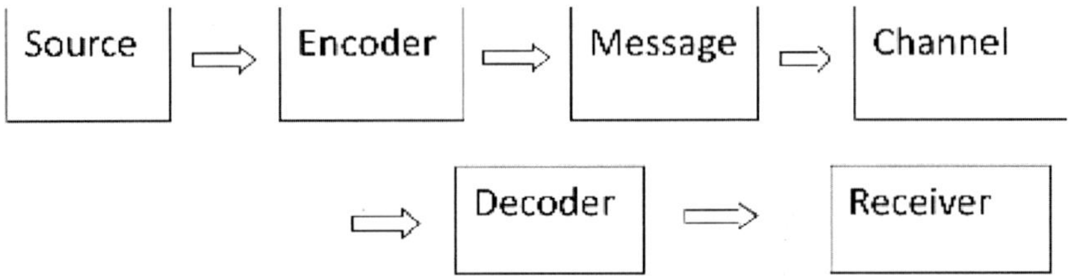

- Rogers and Shoemakers Model

Problems in Communication

- Insufficient Information
- Information Overload
- Over confidence
- Language or Code
- Not Listening
- Too ahead of Audience Understanding
- Lack of Empathy
- Ignoring the leaders
- Prejudice
- Beliefs
- Physical Environment
- Disorganized Communication
- Egoism, Over - Confidence
- Perceptional Difference

Types of Communication

- Formal Communication
- Informal Communication

According to Direction of Communication

- Vertical
- Lateral
- Diagonal

According to expression and body language

- Oral
- Written
- Non Verbal

Essentials of Sender/ Communicator

7 C'S

1. Credibility - Builds Trust
2. Courtesy - Improves Relationship
3. Clarity - Makes Comprehension Easier
4. Correctness - Builds Confidence
5. Consistency - Introduces Stability
6. Concreteness - Reinforces Confidence
7. Consciousness - Saves Time

4 S'S

1. Shortness - Economizes
2. Simplicity - Impresses
3. Strength - Convinces
4. Sincerity - Appeals

Diffusion versus Dissemination

Adoption

- It is a decision making process.
- Decision to make full use of an innovation as the best course of action available.

Steps of Adoption Process

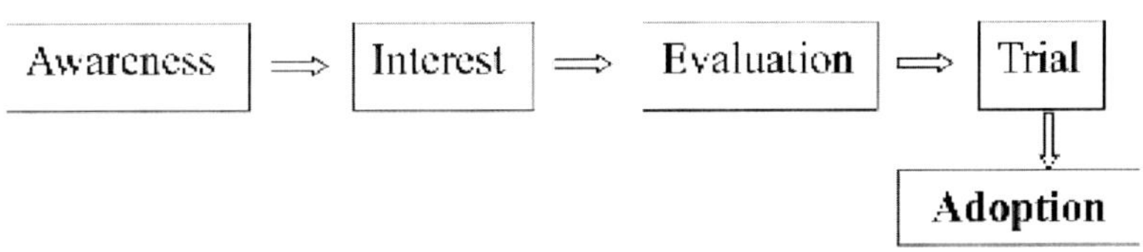

Attitude

- It is a mental state towards extent of readiness to any idea, practice or object.
- It is two type:
  1. Negative
  2. Positive

Adopters Categories

1. Innovators
2. Early Adopters
3. Early Majority
4. Late Majority
5. Laggards

Strategic Research and Extension Plan (SREP)

The strategic research and extension plan basically has two sections namely diagnostic section and strategic section. SREP is a document showing our future plans and procedure or techniques to achieve our goal of increasing agricultural production and farmer's income. This plan is down-top instead of top-down in nature. SREP has been prepared with active participation of farmers, NGOs, field officers and scientists.

❑❑❑

Care and Management of Livestock During Disasters

Natural disasters are beyond human control that cannot be stopped but by taking precautions may either evade such disasters or at least lessen their effects, whereas manmade disasters are manageable. Different types of disasters such as earthquakes, earthquakes (Tsunamis), cyclones, floods, drought, landslide & avalanches, hurricanes, tornadoes, blizzards, Ice storms, wildfires, chemical or nuclear can occur any time. Disasters weather natural or manmade causes great loss to all sorts of life -human beings, animals, plants and resourses-buildings, infrastructure and above all causes physiological problems.

Handling disasters, those catastrophic events which stretch the capacity of communities, can only be approached with preparedness, pre-planned reaction and post-event mitigation. During a disaster event, rural residents often find their personal safety a large enough challenge without the added burden of caring for livestock. In many instances, the response time and resources are greatly reduced. India is vulnerable in varying degrees to a large number of natural as well as man-made disasters. 58.6 per cent of the landmass is prone to earthquakes of moderate to very high intensity; over 40 million hectares (12 per cent of land) is prone to floods and river erosion; of the 7,516 km long coastline, close to 5,700 km is prone to cyclones and tsunamis; 68 per cent of the cultivable area is vulnerable to drought and hilly areas are at risk from landslides and avalanches. Some basic realities of care and management of livestock during disasters are discussed in this chapter.

Earthquakes: Earthquakes the most feared of all natural hazards as they occur without any recognizable warning are unpredictable in space and time and inflict heavy losses in less than a minute duration. About 50-60 percent of total area of our country is vulnerable to seismic activity of varying intensities. Most of the vulnerable area is generally located in Himalayan and sub-Himalayan regions and in Andaman and Nicobar Islands. During Earthquakes livestock like Cattle, Buffalo, Sheep, Goat, Dogs, Pig and Poultry population should be evacuated and shifted to right place immediately. Essential medicines, vaccines and equipments are to be shifted to safer places for proper and quick treatment to suffered animals. The livestock holders of large animals, small animals and poultry should be trained regarding shifting of animals/ birds before disasters. Shifting of poultry birds to safer places should be done made up of suitable cages from bamboos.

Cyclone: Cyclone is a violent storm, often of vast extent characterized by high winds rotating about a calm center of low atmospheric pressure. This center moves onwards, often with the velocity of 50 km an hour. In India they generally originate in the Bay of Bengal or the Arabian Sea during the pre monsoon (April and May) and post monsoon (October to December) seasons. The affected injured / sick animals and birds are to be treated immediately as possible as per the requirement. The existing inventory of medicines is to be utilized.

Floods

Floods occur when large volume of water from heavy rain fall and /or river spill is not able to drain-off quickly through normal channels and are the most frequent and most wide spread natural disasters resulting into death, destruction, degradation and displacement.

Flash floods defined as floods which occur within six hours of the beginning of heavy rainfall and are usually associated with cloud burst, storms and cyclones requiring rapid localized warning and immediate response if damage is to be mitigated. Flash floods usually occur in areas where the landscape cannot absorb all the water during excessive precipitation events in steep gradient landscapes. Living downstream from a significant precipitation event is a danger zone. Livestock have a natural move away instinct to flash flood waters. They generally seek higher ground if possible. When purchasing or designing your livestock operation, it is important to allow livestock a way to reach high ground in each pasture. Without access, livestock will fight fences and be at a greater risk of drowning. Livestock will initially panic during flash floods. This complicates livestock handling and management. Since large-scale floods are more predictable and usually slower to develop than flash floods and you may have more response time. Livestock will move gradually away from rising floodwaters to higher ground. Much before the rush of flood water to flood prone areas, the livestock holders of cattle, buffalo, sheep, goat, pig etc should be informed well in time to let loose their domesticated animals and birds instead tieing/ caging at a particular point. Necessary arrangements should be made for provision of life saving boats for rescue of marooned animals, transportation of feeds, medicines and vaccines. Sincere efforts are to be done in this regard to save the lives of the livestock and poultry. The small ruminants like sheep and goat are carefully shifted to temporary sheds located in higher elevations. Similarly, the poultry birds are shifted with the help of bamboo cages to temporary pen.

Landslide & Avalanches

Landslide constitute a major natural hazard in our country which accounts for considerable loss of life and damage to communication routes, human settlements, agricultural fields and forest lands. Sudden downslides of ice, snow and dirt are difficult to plan against. Moving livestock away from those areas when risks are present is highly recommended. There is seldom time for care, management or animal reaction during these events.

Drought

Drought is a silent disaster (along with famine and pestilence) because it has a slow onset period that does not encourage monitoring because of its discomfort. The key elements to managing animals in drought disaster are food, water and shelter. Provision of immediate clean, uncontaminated water to the livestock. The lack of any of these factors or a scarcity of one or more, can lead to a slow death for livestock. Animal reaction to drought is slow and vague until at critical health levels. Of all the disaster types, drought has the greatest potential to affect the widest area, often impacting multi-state regions at the same time.

Tornadoes

Tornadoes have extreme intensity wind speeds 2 to 3 times that of a hurricane but they have a very short duration. Livestock hear and sense impending tornadoes. If your personal safety is not an issue you may only have time to open routes of escape for your livestock. Do not cut your safety margin short since tornadoes can veer change speed and change footprint width very quickly. Open the gate if you must and then exit the area in a tangent direction away from the expected path of the twister.

Hurricanes

Hurricanes usually become an issue for livestock when the storm reaches landfall. Hurricanes often generate coastal flooding and are accompanied with high winds and heavy rain. Advance warning is usually available so livestock can be moved inland prior to the storm's arrival.

Blizzards

Snow blizzards can easily reach disaster status if conditions are right. Most livestock move away from the storm onslaught unless they are moving to a habitual source of shelter. Some livestock shelters (canyons, draws and windbreaks) can start out as protection in a snow storm and quickly become "drift over" hazards, so you may need to move livestock to shelter areas that have reduced risk of being buried by the drifting snow. Livestock will resist being moved from an area with limited protection and usually also resist efforts to move them into the face of a storm. Young animals are a special risk since they can get buried in snow more easily and have less physical strength and less resistance to cold exposure. Address the young animals first and the older livestock will often follow from both maternal and herd instinct. Although animals rarely get frantic or panicky during blizzards they do get determined to avoid the wind cold and poor footing.

Ice Storms

Rapid onsets of freezing rain combined with the risks of blizzards increase the chances for extreme hypothermia. Moving of livestock promptly to shelter where feed is available. Forage is often temporarily inaccessible during and immediately after ice storms. Animal reactions to ice storms are similar to that of blizzards.

Wildfires

Although the surface speeds of wildfires vary from 5 to 40 miles per hour all wildfires generate smoke heat and sound. Livestock are very sensitive and responsive to wildfire anywhere within their sensory range. Normal reactions vary from nervousness to panic to aggressive and resistive escape attempts. Livestock are often injured or killed by fleeing from a wildfire into fences barriers and other fire risks. Once the flight syndrome kicks in it is retained long after the smoke, heat and noise stimuli are removed. Some animal species such as alpacas, llamas and especially horses become virtually unmanageable in the face of oncoming wildfire. In situations like this, experienced handlers (as many as possible), proper equipment and a firm and prompt evacuation approach should be needed. If time is limited because of fire ground speed, open possible escape routes and recapture animals later. In the case of a fast moving fire some landowners spray paint their phone numbers on the sides of livestock before setting them free. Others attach identification tags to animals. If you choose to leave a halter on your animal consider attaching identification such as a luggage tag. Firefighters may cut fences and open gates if time and safety concerns allow. Any deceased animals must be disposed of safely. Human health concerns, workplace health and safety precautions and environmental impacts must be considered. Health camps should be organized in disaster affected areas to restore the normal breeding capability of breedable population as well as to restore the normal health of livestock and poultry.

□□□

Management and Control of Infectious Diseases in Livestock During Flood Disaster

- Flood- Recurrent phenomena in India.
- Area prone to flood- 46.56 mha.
- Occurrence of heavy floods in the country with a marked periodicity.
- Average area affected by flood- 7.51 mha/year.
- Monetary loss- Rs 1800 crore.

Impact of Flood on Livestock

- Inadequate feed and fodder supply.
- Accidental death.
- Occurrence of infectious disease.
- *In flood Affected Areas*
    1. Favorable conditions for propagation of pathogens (bacteria, viruses, parasites, fungi etc).
    2. Infection to livestock through contaminated feed and water.
    3. Occurrences of infectious diseases, some of them are zoonotic disease.

Factors Responsible for Infectious Disease

- *Population Density*: By closer contact with the animals during flood.
- Ecological Changes that favour breeding of vectors.
- Disruption and contamination of water supply.
- Adverse climate and deficiency of micronutrients.

Commonly occuring Infectious Diseases during Flood

Bacterial Diseases

Haemorrhagic Septicaemia (HS) or Pasteurollosis:

- Cattle and buffaloes mostly affected.
- Occurs in swine, sheep and goats too.
- *Causal agents* in Cattle & Buffaloes: *Pasteurella multocida* type I or B, occasionally type 6:E or E:2.
- *P. haemolytica,* type A:3 causes pasteurollosis in goats.
- *Clinical symptoms* - high fever, marked salivation, increased respiration, swelling in head and neck region.
- *Treatment* - antibiotics at the early stage such as tetracycline + sulfonamide.
- *Control* - inactivated vaccine, oil adjuvant vaccine in May-June.
- Disposal of carcass - deep burial (6 ft ditch) using lime or bleaching powder.

Black Quarter (BQ)

- *Causal agent* - *Clostridium chauvoei* - soil borne organism.
- Cattle between age groups - 6 months to 2 years affected.
- *Clinical signs* - characteristic oedemators and crepitating swellings in the hip, shoulder, chest, neck and back.
- *Treatment* - antibiotics such as penicillin, ampicillin, cloxacillin, amoxycillin.
- *For control* – inactivated vaccine in April-May.
- Combined HS-BQ vaccine.
- Deep burial of carcass as in HS.

Botulism

- *Causal agent* – neurotoxin produced by *Clostridium botulinum.*
- *Occurrence* – during prolonged drought or flood.
- Inadequate feed supply, reduced phosphorus intake, manifested by pica predisposes to infection.
- *Clinical signs* – severe case: early muscle tremors, motor paralysis leading to recumbency.

- *Treatment* - antitoxin in early stage.
- *Control* - avoid exposure to toxins and provision of mineral supplements.

Anthrax

- Fatal infectious and zoonotic disease caused by –*Bacillus anthracis.*
- Characterized by septicaemia, sudden death, exudation of tarry blood.
- Failure of blood to clot, absence of rigor mortis and spleno megaly are important findings.
- *B. anthracis*–a possible agent for bioterrorism.
- Spores formed when organisms exposed to air and is capable of infecting both humans and animals.
- Spores survive in soil for longer period –10 yrs or upto 60 yrs.
- During flood spores in soil are surfaced due to soil erosion.
- Infection to man and animals through contaminated feed and water.
- Disease affects -cattle, sheep, goats (horses, pigs, camel, elephants may also suffer).
- In per acute form –death within 1-2 hrs.
- *Symptoms* –momentarily fever, staggering gait, muscle tremors, dyspnoea and collapse.
- Discharge of blood from nostrils, mouth, anus and vulva.
- In acute form –fever, excitement, depression, difficulty in respiration, staggering, convulsions, stupor.
- Duration in acute form –24-48 hrs.
- *Treatment* –antibiotics, if done at early stage.
- Immunization with anthrax spore vaccine before rains or expected flood.
- PM examination is contraindicated.
- Disposal-same as in HS.

Leptospirosis

- Water borne zoonotic disease.
- **Characteristics** - septicaemia, interstitial nephritis, haemolytic anaemia and abortion.
- **Caused by** different serovars of *Leptospira interrogans.*
- Cattle, buffaloes, sheep, goats and pigs suffer.
- Rodents play pivotal role in disease cycle.
- Occurrence of disease more during monsoon and flood.
- Infected animals excrete leptospira organisms particularly in urine.
- Organisms remain viable in river, pond, canal and mud and contaminate the water.
- During flood the susceptibility of the disease is increased due to the flow of water.
- **Symptoms** - fever, jaundice, abortion, haemoglobinuria and nephritis.
- **Treatment**- antibiotics: dihydrostreptomycin, oxytetracycline, ciprofloxacin, entrofloxacin.
- No vaccine is available.
- **Prevention** - animals should be kept away from contaminated water and control of rodents (rats) in human dwellings to prevent the spread of infection.

Viral Diseases

Foot and Mouth Disease (FMD).

- Highly contagious and has economic importance.
- Economic loss > Rs 25,000 crore.
- Loss of milk production, poor reproductive performance, endocrine dysfunctions.
- Virus affects host immune system and makes animals susceptible to several infections particularly haemoprotozoan diseases.
- **Causal agent**–FMD virus, types: O, A, C and Asia-1.
- In Bihar and other states 70% of Out Breaks by type 'O' only.
- Out Breaks are observed during flood.

- **Control** - regular vaccination programme.
- Primary vaccination at 4 months.
- Revaccination at 9 months.
- Then, subsequent revaccination annually.
- In high risk areas -vaccination at 6 months interval.
- Mass vaccination under field condition: Jan-Feb.

Rabies

- Oldest fatal viral zoonosis.
- Affects all warm blooded animals.
- Prevalent in dogs, cats, wild carnivores: jackals, foxes, wolves and mongoose.
- **Carriers** -wild carnivores.
- During flood carriers enter into human habitat increasing possibility of transmission.
- Untreatable, prophylactic or post exposure vaccination is required for protection.
- After bite of rabid dog: thorough washing of wound with soap tepid water for 15-20 minutes.
- Application of Tr. Iodine or Betadiene on wound.
- Post bite rabies vaccination.
- Immunization of pet dogs regularly.

Gastro Intestinal (GI) Parasitism

- High morbidity and mortality in livestock.
- **Symptoms**–digestive disorders, reduced body growth in young animals, poor productivity in adults.
- Snail borne parasitic infections: *liverfluke* and *amphistomiasis* common in flood affected areas.
- Cattle, buffaloes, sheep and goats are affected.
- Intermediate hosts i.e. snails of *Lymnoea* and *Indoplanorbis* become abundant in flood affected areas.
- **Control**–adoption of regular deworming schedule.

Deworming schedule

- Young calves should be dewormed against *ascariasis* at the age of 2-3 weeks.
- Repeated dewormings at the interval of 1.5-2 months till the age of 5-6 months.
- Anthelmintics used consist of *Morental Citrate, Fenbendazole, Levamisole* and *Tetramisole* in required doses.
- Anthelmintics should be used by rotation.
- For other nematode infections in growing calves deworming required at 2-3 months till 2 yrs of age.
- Above mentioned anthelmintics are to be used in growing calves.
- Deworming against *liverfluke* and *amphistomiasis* in adults should be done 3-4 times.
- 1st deworming : Jan-Feb, 2nd : Apr-May, 3rd : Jul-Aug, 4th : Oct-Nov.
- Anthelmintics used: *Oxyclozanide, Albendazole* and *Rafoxanide* in proper doses.
- For adults, deworming should be done at least twice with broad spectrum anthelmintics against the mixed infection of nematodes and trematodes.
- Broad spectrum anthelmintics Combination: Oxyclozanide + Levamisole or Oxyclozanide + Tetramisole or Albendazole alone.
- All the anthelmintics are also effective in case of sheep and goats and they should be dewormed thrice in a year following the same schedule.

Haemoparasitic Diseases

- During monsoon and flood, the activity of the insect vectors (ticks, flies etc) becomes enhanced due to favourable conditions for their propagation.
- They become capable of transmitting the haemoprotozoan diseases particularly *Theileriosis* and *Trypanosomiosis* both during and after flood.

Theileriosis

- **Causal agent**–*Theileria annulata*
- **Transmitted** by *Hyalomma* ticks
- Cross bred cattle of all age groups are susceptible
- Severity of disease is high in young calves with high mortality rate
- **Treatment**–*Buparvaquone* is available in market by different trade names
- **Control Measures**:
  1. Control of vectors
  2. Immunization of susceptible animals
- For control of vector ticks ectoparasiticides: *Delta methrine*or *Cyper methrine*is used
- For immunization: Schizont vaccine has been developed which is lymphoid cells grown schizonts attenuated after passages
- Vaccine is not popular due to certain limitations
- Not recommended for immunizing young calves below 2 months of age.

Trypanosomiosis

- Widely spread in rural areas of Bihar and other states
- **Occurrence**–during monsoon and flood
- **Caused by**–Trypansomaevansi and transmitted by *Tebanus* fly
- **Treatment**–Quinapyramine sulphate& chlorideand Diminazine diaceturate
- No vaccine available for control
- In high risk areas it can be controlled by chemoprophylaxis
- Cattle and buffaloes are given one injection of *Quinapyramine sulphate*and *chloride*combination before monsoon
- This affords protection for 3 months.

Degnala Disease (Mycotoxicosis)

- In flood prone areas, the stored fodder (paddy straw) becomes moist and mouldy.

- **Growth of fungus**-*Fusarium equisetti* which produces mycotoxin.
- After feeding such paddy straw, the animals develop symptoms of toxicity.
- **Clinical symptoms**–lameness, swelling of lower parts of legs followed by oedema, necrosis.
- Gangrene of extremities (in lower parts of legs, tail and ear lobes) develops in advanced stage of disease.
- Sloughing of skin and complete separation of hooves occurs in severe involvement of feet.
- **Treatment**–injection of analgesic, anti-inflammatory, anti histaminic and anti biotic drugs.
- Supported with injection of *Vit-A, Skin tonic (Acetylarson 23.6% solution)* and feeding mineral mixture.
- **Prevention**
  1. Paddy straw should be stacked at high level.
  2. It should be sufficiently dried before stacking.
  3. Moist and mouldy straw should not be fed to animals.
  4. Mineral mixture should be given regularly as feed supplement.

Disaster Risk Reduction Strategies During Draught in Livestock and Birds

Draught is a climatic anomaly, characterized by reduced supply of moisture either due to sub-normal rainfall, erratic rainfall distribution, higher water need or a combination of all these factors. Droughts differ from other natural disasters in several ways. It is a slow-onset, creeping phenomenon that makes it difficult to determine the onset and end of the event and hence duration may range from months to years. There is no single indicator or index that can identify precisely the onset and severity of the draught. Likewise, impacts are generally non-structural and difficult to quantify. Spatial distribution is also very large making assessment and response actions difficult. There is no universally acceptable definition of draught. Meaning of draught varies for people belonging to different profession. For a meteorologist, it is the absence of the rain, while for the agriculturist; it is the deficiency of soil moisture in the crop zone. For a hydrologist, it is the lowering of water levels in lakes, reservoirs or depletion of ground water; while for the city management it may be the shortage of drinking water supply. Draught proceeds in sequential manner and its impacts spread across different domains as follows:

a. *Meterological draught*: It is simple deficit/absence of rainfall from the normal. It is the least severe form of draught and is often identified by sunny days and hot weather.

b. *Hydrological draught*: Hydrological draught may lead to reduction in natural stream flows, river water and groundwater levels. All these results into shortage of water resources for human and animal consumption.

c. *Agricultural draught*: It results when soil moisture is insufficient to maintain average crop yields. In less severe form, it results into reduced output of crops and other agricultural products. In sever from; it can lead to a famine.

d. *Socioeconomic draught*: It correlates the supply and demand of goods and services with the three above-mentioned types of draught. During draught, shortage in supply of water and electricity may occur.

Draught prone areas in India: In India draught-prone areas comprise a total land area of 329 million hectares, with three-fourths being arid, semi-arid and sub-humid areas:

1. *Arid Zone (19.6%):* Mean annual precipitation (MAP) of 100-400 mm. It includes Rajasthan, parts of Haryana and Gujarat. Draughts are severe in this zone.
2. *Semi-arid zone (37.0%):* MAP of 400-600 mm. Include parts of Haryana, Punjab, west Uttar Pradesh, West Madhya Pradesh and peninsular parts of Western Ghats. Draught can be moderate to severe in this zone.
3. *Dry sub-humid zone (21.0%):* MAP of 600-900 mm. Include parts of northern plains, central highlands, eastern plateau parts of Eastern Ghats and plains and parts of western Himalayas. Draught is moderate in this zone.
4. Humid and pre-humid regions, such as Assam and other north-east States rarely face draught.

Impacts of draught: Impacts of draught are felt much beyond the geographical area experiencing physical draught. Impacts of draught are both direct and indirect. Direct impacts or primary impacts occur in terms of physical/material losses and include reduced agricultural production, increased fire hazard, depleted water levels, higher livestock and wildlife mortality rates and damage to the environment and wildlife. Indirect effects include reduced agricultural production, reduced farmer?s income, increased prices of food, unemployment, reduced purchasing capacity and demand for consumption. Social impacts of draught include migration of people from draught affected areas, sale of assets like land or livestock and decrease in social dignity.

Preparing for draught: Draught is a slow-onset phenomenon. Therefore, farmers may be guided to prepare in advance to fight against drought. This can help in reducing losses occurring due to draught.

Draught mitigation: Role of Veterinarians: A veterinarian is mainly concerned regarding health care and management of animals during draught. Guiding farmers and common people regarding efficient management of available water, feed and fodder resources and issues related to animal welfare also come under his preview. If danger of draught is impending following stepts may be taken by the livestock farmers to reduce the economic losses:

1. Adjustment in herd strength

Prices of hay, green fodder and grains increases during draught. Sometimes green fodder, hay and water may be available only in limited quantity. Therefore, it is very important to retain only those animals that can be provided food and water during the draught. Assess the cost and benefit ratio for all animals and retain only those which are best fitted in terms of long term economic return. Animals should never be left to starve to death or die of thirst. Three options can be adopted which include, 1. Retaining 2. Agistment and 3.Selling. The best animals to retain include young breeders in the age group of 2 to 5 years with good genetic potential, because it will be difficult and expensive to purchase such animals in future for replacement. Females should be tested for pregnancy and sell all empty, late calving and low producing animals. Inferior bulls, aged/ poor performing cows and bulls and young or weak male calves or heifers should also be disposed by selling. Usually prices of animals during draught decreases drastically and hence it may not be desirable to sell animals. One can wait to increase the prices. Sending animals away on agistment may be helpful in such conditions. Shift animals to places where water, feed and fodder are available in adequate quantity at cheaper rates. But, only strong and healthy animals able to withstand the journey stress or able to walk long distance should be agisted. At suitable time, either sell or return the agisted animals to join the parent herd.

Storage of water, feed, fodder and concentrate

Farmers should be advised to procure and store hay and feed in quantity sufficient to meet their requirement till draught ends. If green fodder is available, silage can be prepared for use in future. Poor quality forages can be ammoniated to increase to nutritive value. People should be advised to prepare artificial catchments for harvesting rain water. The stored water can be used in future for various purposes in livestock husbandry.

2. Nutrition and feeding

Management of feeding practices to ensure proper nutrient uptake during draught is essential for maintaining health and getting optimum production from animals. Availability of good quality forage is important for obtaining optimum milk production and normal milk composition and quality, as well as, maintenance of health and reproductive potential. Supply of good quality fodder crops is often difficult during draught. Draught affected fodder crops may show stunted growth, fail to reach flowering stage or develop grains and usually have low moisture content. They are, in general, low in vitamin A, phosphorus and protein content, but have high potassium content. Therefore, supplementation of deficient nutrients is essential in cattle being fed on fodder, hay or silage produced from such plants. Important nutrients or feed supplements required during draught are as follows:

A. *Phosphorus*: One can continue feeding salt and mineral mixture that was being fed during normal conditions. However, to meet increased phosphorus demand, a mixture of trace mineralized salt and dicalcium phosphate (in ration 50:50) would be better to use. If possible reduce the salt content so that water consumption by animals is reduced.

B. *Vitamin A*: Vitamin A concentration in forages grown during drought and hay produced from drought-affected forages are very low. Therefore, supplementation of vitamin A is crucial particularly in lactating and pregnant animals. Booster doses of vitamin A and D can be given to pregnant animals 30 days prior to calving. The newborn calves should also be supplemented with vitamin A and D at regular intervals.

C. *Protein*: Since protein content of stunted fodder plants or grass is low, its supplementation may be essential particularly in breeding and lactating animals. Protein deficiency increases infertility. Dry cows should be given 225 to 340 grams of supplemental crude protein, while lactating cows should receive about 400 to 540 grams supplemental crude protein per day. Soybean meal is an easily available good quality proteins source. Therefore, soybean meal can be given 900 to 1150 grams to lactating cows and 450 to 670 grams to dry cows, per day. Other protein rich sources include Alfalfa hay, Sunflower meal, Berseem, pulses, oil-seed cakes etc.

D. *Energy*: The ability of the animal to maintain their weight and production level (growth, reproduction and lactation) is highly dependent on meeting specific energy requirements. Energy may be a limiting factor in animal fed entirely on poor quality roughages and fodders grown during draught. Low quality forages can be ammoniated to increase digestibility and protein content. Other energy rich sources include grains (e.g. corn, oat, barley etc.), molasses, hay and crop and grain processing byproducts (beet pulp, barley malt sprouts, corn gluten feed etc).

E. Water Adequate availability of good quality water is often a problem during draught. Water is essential of animal survival and performance. The requirement of water for animals depends upon following factors:

   a. *Physiological state*: A lactating cow requires more water than a dry cow of same body weight.

   b. *Environmental conditions*: Rise in surrounding temperature increases water requirement.

   c. *Type of feeding material*: Animals on dry feed requires more water, while those offered green fodder may require less water.

   d. *Quality of water*: Hard water (with more salt) is consumed more.

Economical use of available water should always be ensured. Farmers may be advised to use good quality water only for drinking purposes. For other livestock related works, like those required for sanitation etc., they can use non-potable water. In draught prone areas, farmers may be advised to built rainwater harvesting tanks in which water collected on the roof of the house can be stored for future use.

3. Disease management

A. *Draught related poisonings*: Draught related poisoning occurs due to consumption of poisonous plants. There are two possible reasons. In some cases, it occurs due to grazing of poisonous plants that are normally not consumed by animals. But, hungry animals in overgrazed areas, when desirable grasses and forage are limited, may consume relatively unpalatable poisonous forages. However, certain forages and grasses that are normally consumed by animals without any harmful effects, when grown under stress or limited water supply, accumulate certain toxic metabolites, like nitrate, hydrocyanic acid (or prussic acid), prussic acid etc. Therefore, chances of fodder/ forage poisoning increase when hot, dry weather conditions persists. Some important poisonings occurring commonly during draught conditions are as follows:

i. *Lantana poisoning*: *Lantana camara* is a multi-branched thicket-forming shrub, found throughout India. They can survive well during draught and contain toxic triterpene acids, lantadene A (rehmannic acid), lantadene B and their reduced forms. Grazing animals generally do not consume it because of its pungent aroma and taste. But, it may be consumed by animals during draught, or by animals unfamiliar to the plant. Toxicity symptoms develops when an animal consume it in quantity more than 1 per cent of the animal's body weight. Species affected include cattle, buffalo, sheep and goat. Clinical signs of toxicity may appear within 24 hours after its consumption and include anorexia, depression, weakness, frequent urination, constipation, photosensitization (excess sensitivity of skin to sunlight) and jaundice. Muzzle becomes inflamed, moist and ulcerated and unpigmented skin and tip of ears may slough in some cases. Diarrhea with strong-smelling black fluid faeces may develop in severely affected animals. Affected animals should be kept in shade and drenched with slurry of 2.5 kg activated charcoal. A second dose of activated charcoal may be required after 24 hours. Plenty of fluid and balanced electrolyte solution should be infused and liver tonics should be given regularly. Enlarged and discoloured live with a yellow to orange colour and very swollen gall bladder are characteristic postmortem findings in affected animals.

ii. *Nitrate poisoning*: Extremely dry or cool weather conditions may prevent plants form converting nitrate to true protein. Therefore, nitrate may accumulate in excess in stressed forages particularly in Sorghum and Sudangrass. Excess use of nitrogen fertilizers without proper supplementation of phosphorus and potassium may further increase the nitrate content in plants. Nitrate level decreases during ensiling because it is converted into nitrogen oxide gas and is evaporated. Plants suspected to contain excess nitrate should be cut higher above the ground than usual, as nitrate tends to accumulate in the stem. It is better to use such fodder crops for silage making rather than using as green fodder. If suspected green fodder is being given to animals, introduce it gradually and also give about 1.5 kg concentrate per head per day. Nitrate concentration below one percent on dry matter basis is safe. Nitrate toxicity occurs when animals consume large amount of forage containing 2-3 per cent nitrate or more on dry matter basis. Clinical symptoms of toxicity include bluish mucous membrane, rumen paralysis, laboured breathing and sudden death. Sometimes lower levels of nitrate may cause subacute poisoning, where milk production and appetite is generally not affected, but reproductive problems like abortions may occur. Level of nitrate in fodder as well as drinking water should be checked because deep bore well water may be another source of excess nitrate. Intravenous administration of 4 percent methylene blue in 5 percent dextrose is recommended in acute toxicity cases.

iii. *Silo gas poisoning*: It occurs due to excess production of nitrogen oxide gas by improper fermentation of plants containing high nitrate concentration. Problem occurs commonly while ensilaging draught affected plants with low moisture content. The gas is highly toxic to animals, birds and humans. Inhalation of nitrogen oxide causes choking, coughing and burning sensations in nostrils, throat and chest. Affected animals show coughing and choking. Accumulation of fluid in chest and lung damage occurs. To avoid the problem, silos should be well ventilated and one should not enter silos for at least 7-21 days after filling with suspected forage. Plants with moisture level around 60-70 percent should be used for silage preparation.

iv. *Prussic acid poisoning*: It occurs due to intake of immature Sorghum and Sudangrass that are grown under severe stress like drought or frost. Symptoms resemble those observed in cyanide poisoning. In acute toxicity cases animals show lack of coordination and convulsions followed by quick death. Treatment with sodium thiosulfate (@66 mg/ kg. b.wt.) and sodium nitrite (@22 mg/kg. b.wt.) mixed in 200 ml distilled water, given as slow I/V is recommended.

v. *Mycotoxins*: They are more likely to be present in corn and grains than fodder crops. However, improperly prepared silages and hays may contain certain mycotoxins like aflatoxin and zeralenone. Cattle affected with aflatoxin passes black or bloody diarrhea. Milk production decreases and posterior paralysis may occur. Aflatoxin damages liver. Cattle affected with xeralenone shows swollen vulvas and abortions.

B. *Draught related diseases and problems*: In general, starved or poorly fed animals become more prone for various diseases. Therefore, special care should be taken to avoid disease outbreaks. Animals should be vaccinated regularly. Sick animals must be supplemented additional minerals and vitamins to strengthen their immune system and ensure fast recovery.

   i. *Heat stress/ heat stroke*: Problem of hyperthermia or heat stress increases many fold during warm windy weather. Animals should be housed in shaded and well ventilated areas. Avoid movement or taking any kind of work from animals during hot hours of the day. Adequate water supply can minimize the effects of heat stress.

   ii. *Metabolic diseases and deficiency diseases*: Incidence of different metabolic diseases increases during draught due to improper nutrition. Milk fever, post parturient haemoglobinurea and ketosis are important diseases for which animals should be examined regularly. Anemia, weakness and cachexia may develop in non-producing dry animals. Reproductive performance of both male and female animals is severely affected. Hypomagnesemia may occur in calf fed entirely on milk with little or no forages, grains or mineral mixture. Magnesium content in young forage plants and grasses grown during winter or in old forages is usually very low. High potassium content of such crops further prevents magnesium absorption. Therefore, problem occurs more frequently during winter months. The affected calf lies down, unable to get up, shows tetany or convulsions and may die. Cows in early lactation show nervous signs. In such cows, mineral mixtures containing 10 to 15 percent magnesium should be supplemented on regular basis.

   iii. *Infectious diseases spreading from silage*: If silage fails to develop a pH around 4.8-5.0, chances to spread entertoxemia, botulism, listeriosis and mycotoxcosis increases. This can happen due to either using plants with low moisture content for ensilaging or adding urea or anhydrous ammonia or other NPN sources during ensilaging. Faulty ensilaging procedure or leakage in silo pits may be other possibilities resulting into improper fermentation and high pH.

iv. *Other diseases*: Weak animals become more prone for various infectious diseases due to compromise in immune system. Incidence of diseases that spreads through wind movement increases may fold during draught. Important diseases under this category are FMD and Flu. Adoption of proper vaccination program can reduce occurrence of such diseases in animals. Prevalence of tick infestation and tick born diseases also increases when dry warm climate persists for long. Care should be taken to reduce the tick infestation in domestic animals. In ruminants, problem of impaction increases due to low water intake and less/ no green fodder availability. Animals may suffer from inanition and loose body weight. Ensure proper water intake, specially when dry fodder is given as the major diet or green fodder is not being served.

❑❑❑

Chapter 14

Impact of Natural Disaster on Wild Birds

Unnatural disasters like oil spills and airplane strikes generate a lot of interest in bird conservation, but in fact natural disasters may be equally destructive to wild birds. In many cases, a large scale natural disaster can be even more harmful than an unnatural event because not only it is detrimental, but its effect is relatively overlooked and the affected birds may receive little assistance.

Natural Disasters that affect Birds

All types of natural disasters can have severe impacts on wild birds, and even events that may not seem entirely disastrous can have grave consequences depending on how, when and where they strike. Disasters most likely to affect birds include:

- Hurricanes, tornadoes and other strong winds
- Thunderstorms
- Blizzards, ice storms and severe cold
- Earthquakes and landslides
- Volcanoes
- Floods and tsunamis
- Forest fires and wildfires
- Droughts and heat waves

The Effects of Natural Disasters on Birds

Depending on the type of natural disaster, one or more devastating effects can impact wild birds in the affected area.

- *Nesting Destruction*: Nesting areas can easily be burned, flooded or buried by natural disasters. If the event occurs during the breeding season, chicks, eggs and brooding adults can be killed, or if the event occurs outside the breeding season, the critical nesting habitat may be unsuitable for the next year's reproduction cycle, which cans greatly impact future population numbers.

- *Food Destruction*: Food sources such as grain, seeds, insects, nectar and small animals can all be destroyed by natural disasters. This leaves birds without adequate food supplies and forces them to expend much more energy in the search for food or competing with other individuals over what little food remains.
- *Habitat Damage*: Natural disasters can damage or destroy the habitat birds need to survive. Whether it is a bird's nesting area or not, the loss of suitable habitat can impact birds by removing shelter and roosting areas that are critical for year-round birds as well as for migratory species.
- *Direct Deaths*: Some natural disasters can directly cause the deaths of otherwise healthy birds. Sudden winds can blow birds into obstacles and cause death by impact trauma, heat waves can cause dehydration and cold snaps can freeze birds that don't have adequate protection.
- *Migration Disruption*: Natural disasters can devastate birds if the event interferes withmigration. A flight of hundreds or thousands of miles is already hazardous, and storms that require detours can exhaust birds and create much higher migration mortality. Furthermore, birds forced to migrate along an unfamiliar route may run into additional hazards, such as lower food supplies or unexpected predators.
- *Creating Unnatural Disasters*: Natural disasters have the possibility of creating unnatural disasters that can cripple birds. A hurricane, for example, could damage an offshore oil rig and cause an oil spill that would affect birds and other wildlife long after the storm has dissipated.
- *Less Assistance*: In the wake of a natural disaster, wild birds suffer from lowered interest in conservation efforts as donations are redirected toward emergency relief services. In the long term, this can impact birds by decreasing the overall effectiveness of conservation initiatives.

Birds Adaptation against Natural Disasters

Cumulatively, the effects of natural disasters on birds may seem horrendous to the point where it is a miracle that any birds survive at all. While some species are more heavily impacted by these events than others, many types of birds are resilient and can adapt to natural events in different ways, such as:

- Changing ranges or territories to more suitable areas over time.
- Adapting to a niche that benefit from certain disasters, such as feeding on insects in dead trees after a forest fire.
- Irrupting to take advantage of different territories or food sources until the impacted area recovers.
- Raising multiple broods of chicks to replace young that may be lost to a disaster.

Despite these adaptations, however, natural disasters still devastate many types of birds, and proper conservation is essential to help species recover.

❑❑❑

Disaster Risk Reduction Strategies During Outbreaks of Zoonotic Diseases in Livestock and Birds

Anthrax

Anthrax is caused by a bacterium, *Bacillus anthracis*. When exposed to the air, the bacteria form resistant-spores that can remain viable for at least 50 years when buried in the soil. Anthrax is a zoonosis and people can become ill and die from this disease (anthrax in humans is a notifiable disease). Therefore disease investigations require specific sampling procedures, carcase disposal, site decontamination and careful attention to the safety of all people involved.

Health and safety

In the 1920s and 1930s, anthrax cases were associated with infected shaving brush bristles. A number of people died from this source of infection since there were no antibiotics available at that time. In the early 1960s, a farm worker contracted cutaneous anthrax after conducting autopsies on dead sheep. The worker refused early medical treatment and died from the disease. A veterinarian investigating a disease situation where anthrax is a possible diagnosis must take precautions to avoid potential personal infection or infection of other people. Livestock owners should be advised of the human health risks. All people on the property should be instructed to keep away from the carcase and any contamination from it, unless appropriate personal protective equipment (PPE) is worn and appropriate decontamination conducted afterwards. Anyone who has been exposed to a suspect case of anthrax should seek urgent medical advice. They should contact a doctor or local health unit.

Clinical and autopsy signs of anthrax in animals

The first indication that grazing animals may have anthrax is usually when sudden deaths are reported. If they survive long enough, clinical signs of weakness, staggering and laboured breathing may be observed. Blood may be seen in the urine and milk of affected animals. Pigs affected by anthrax are usually seen to be ill as they tend to survive longer than grazing animals. Dogs and cats are generally resistant to anthrax. After death, blood-stained discharges at external orifices are characteristic of the disease but not all anthrax cases show these signs. There is rapid putrefaction of the carcass and the blood may fail to clot, continuing to ooze from orifices and where predators have attacked the carcass e.g. eyes and skin. If the carcass has been opened, the pathological signs of anthrax are oedema, petechial or ecchymotic haemorrhages on the serosal surfaces and unclotted blood in the large veins. An enlarged haemorrhagic spleen is characteristic of anthrax.

Field investigation

Anthrax is a notifiable disease and any person suspecting anthrax in an animal is required to notify the government immediately. If a dairy farm producing milk for human consumption is involved, the milk distribution should be withheld immediately while investigations proceed. Wherever possible, a veterinary officer will investigate suspect cases of anthrax in animals and conduct autopsies.

A full disease investigation must be conducted and a report submitted. The following must be recorded:

- a full history of the property and animals concerned
- Any cattle that have recently entered the food chain
- All clinical signs shown by any ill animals
- Pathological changes shown by dead animals.

The danger with anthrax is that the bacteria can form spores that will stay in the environment for many decades as a source of infection. The following precautions are essential:

- Do not drive your vehicle close to the carca
- Do not walk in discharges from the carcass
- Do not contaminate your vehicle with carcass material
- Do not open the carcass
- Do not move the carcass.

Equipment

Personal protective equipment (PPE) includes:

- Overalls
- Gloves
- Face mask
- Eye protection
- Rubber boots
- Disinfectant
- Bucket, scrubbing brush etc. for decontamination.

Sampling and packaging materials required per animal are:

- Glass slides
- Scalpel blades
- Syringes
- Blood containers
- A sealable container for blood smears
- Clip-seal bags
- 50 mL containers for soil samples
- Marker pen
- Ice brick for transporting samples.

Disinfectant

Personal disinfection

Prevention of skin contamination by using full PPE is the best option as there are no appropriate disinfectants registered for use on skin. If skin contamination occurs, thorough washing with soap and water will remove anthrax organisms and significantly reduce the possibility of infection. Water used for washing should be contained to a small area and this area must be disinfected with gluteraldehyde or formaldehyde at a later time. A doctor of local health unit should be consulted as a precaution.

Equipment and clothing disinfectants

Gluteraldehyde at 4% (e.g. 200 mL 'Microcide' per 5 L of water) is a suitable disinfectant for clothes and equipment. Exposure should be for four hours. It is not for use on skin. Care needs to be taken in its use. The label directions or Material Safety Data Sheets (MSDS) must be followed. A 5%-formaldehyde solution is suitable for disinfecting equipment. Take care when using formaldehyde or formalin and follow the label's instructions or MSDS.

Conducting or continuing an autopsy

If you suspect anthrax, **DO NOT UNDERTAKE AN AUTOPSY**. In the unopened carcass, anthrax bacilli do not sporulate and are destroyed by the putrefactive process. Collect samples to exclude anthrax. When there is less likelihood of anthrax because of the animal's immediate history or movements, but a suspicion of anthrax still exists because of a history of sudden death and appropriate clinical signs, protective clothing must be worn before proceeding. The carcass should be placed with the near (left) side uppermost and an incision made behind the last rib through the body wall. Lift the rib with a wire hook or similar tool to expose the spleen, viscera and nature of any abdominal fluid. This must be done with care if the carcass is badly bloated to prevent infected material blowing back onto the operator. If an autopsy is carried out in ignorance, and the animal has died from anthrax, the following signs will be seen:

- The carcass meat will look fevered and ooze fluid.
- Blood fails to clot.
- Ecchymotic haemorrhages are present through the body tissues.
- Body cavities contain bloodstained serous fluid.
- There is severe enteritis.
- The spleen is grossly enlarged, soft and the structure liquefied.
- Any subcutaneous swelling are jelly-like and local lymph nodes enlarged. This particularly applies to horses and pigs.

If the carcass is opened, the bacteria form spores on exposure to air and can contaminate the environment around the carcass for many years. If the carcass is already opened when anthrax is first suspected, the autopsy should be ceased immediately and further contamination of the surrounding area prevented. All materials should be confined to a small area for decontamination and destruction. If the PPE being worn is inadequate for anthrax, personal disinfection should be immediately undertaken as set out above and more suitable PPE put on to complete sampling. A doctor of local health unit should be consulted as soon as possible if anthrax is suspected, particularly if a skin injury or needle stick injury has been sustained while conducting the examination and sampling.

Samples

A blood smear and a blood sample are preferred. A blood smear is usually best obtained from the ear or the tail. A small stab is made in the skin with a scalpel blade and a small volume of blood is collected using a syringe. The puncture site should be covered with a swab soaked in a suitable disinfectant. One drop of the blood is used to make a blood smear on a slide which is air-dried and packed securely in a closed container. The remainder of the blood should be place in a secure blood tube. All the samples should be placed in a plastic bag or container and sealed. The outside of the container should be disinfected. If it is not possible to collect from these sites, blood may be extracted from the jugular or brachial veins or as a last resort any other bloody exudates. Any puncture site should be treated with a suitable disinfectant e.g. a cottonwool swab soaked in formalin. If other tissue samples are collected for differential diagnoses, these samples should also be collected and packaged as above. If it is not possible to collect suitable blood samples, environmental (soil) samples should be collected from the ground at the nasal and anal areas of the carcass i.e. where the discharge has contaminated the soil. Around 50 g of soil from each site should be placed in sample containers. These are then placed into plastic bags, sealed and the outside disinfected. Used sampling materials (syringe, tissues, pipettes etc.) should be placed in a container with disinfectant for later disposal. Disinfect the outside of this container and put aside for later removal. All samples must be fully identified and double-bagged.

Transporting samples

The samples above should be placed in a transport box that contains any fluids and materials from breakage of slides or bottles. Decontaminate the outside of the box. Laboratory submission forms should be placed in an envelope and attached to the outside of the box. Contact the laboratory to warn them that suspect anthrax samples are being submitted. As anthrax is a human pathogen, particular care should be taken to ensure that all statutory requirements are met for transporting specimens. Specimens should be packed and labelled for transport in compliance with the carrier's conditions, government and postal regulations and International Air Transport Association (IATA) regulations as appropriate.

Site decontamination and carcass disposal

Animals that have died from anthrax should be disposed off. It is critical to minimise the contamination of the area and the persistence of spores in the soil. It is now known that anthrax spores on the surface of the soil are far more easily destroyed than spores buried below the soil. Burning is the preferred way to dispose of infected carcasses. This should occur where the carcass is situated and as soon as possible. Any people assisting with the operation must be made aware of the risks to human health and the precautions necessary while preparing for burning the carcass. As a final measure, the area of contaminated soil should be decontaminated by the liberal use of a 5% formaldehyde solution at the rate of 5 L/m. It is necessary to wet the soil sufficiently to keep it moist for 12 hours. Care must be taken in using formaldehyde or formalin as per the label instructions or MSDS sheets. If burning is not possible:

- Bury carcasses with a soil-coverage depth of at least 1 m, preferably 2 m.
- Liberally treat the carcass with quicklime (pH 11-13).
- Permanently identify the burial site by recording.
- Decontaminate surface soil using a 5% formaldehyde solution as above.

Where burning or burial is delayed or not appropriate, an alternative approach is to leave carcasses where they are, intact, to undergo natural putrefaction. They must be protected from scavenging by wild animals by an appropriate fence or barrier. Both the carcass and the surrounding soil must be decontaminated with 5% solution of formaldehyde as above. On no account should any animals suspected to have died from anthrax be transported.

Personal decontamination procedure

The routine for personal decontamination should be as follows:

- At all stages of personal decontamination, it is important to avoid contamination of the skin.

- While standing on the contaminated site, gross contamination (mud and grass) should be removed from boots and placed with the carcass.
- Disposable overalls, masks and gloves should be removed and secured with the carcass for later disposal with the carcass.
- Re-usable overalls should be placed in a strong plastic bag or bucket and completely soaked in an appropriate disinfectant for at least 12 hours.
- Boots should be scrubbed with disinfectant; paying attention to remove all material, especially from the soles.
- It is now possible to move away from the contaminated site to clean ground.
- Boots should be placed in a plastic bag or bucket and kept damp for at least 12 hours.
- All reusable protective equipment should be double-bagged for removal.
- Hands and all exposed skin should be washed with liberal amounts of soap and water.
- The site where wash-down has occurred should be marked for later disinfection when the carcass is disposed of. It is essential not to walk back over the potentially contaminated ground in street shoes.
- It is recommended that as soon as possible anyone exposed to the animal or site, have a shower and launder their clothes.

Reporting

When anthrax is confirmed as the cause of death, further actions will be taken by the government to control spread of the disease and further losses. A full report of the disease investigation must be submitted to the Chief Veterinary Officer as soon as possible.

Vaccination

- Vaccination of animals is the key control measure in managing outbreaks of anthrax. A vaccination program will be continued for three years after the outbreak. Vaccinating exposed animals for three years on infected premises following an incident provides time for the dispersal and inactivation of spores on the soil surface to levels unlikely to support further disease incidence.

- Vaccinated animals need to be identified and must not be slaughtered for human consumption for a minimum withholding period of 42 days.
- The anthrax vaccine contains living spores of the non-capsulated naturally avirulent (live) Stern 34F2 strain of *Bacillus anthracis*. A single vaccination is usually effective for 6-12 months, provided that animals receive the full dose and are not under antibiotic therapy within 10-14 days before or after vaccination. Animals that are vaccinated twice, at least six months apart, are probably immune for life.

Glanders and Farcy

Glanders is a fatal infectious disease of horses, donkeys, and mules caused by *Burkholderia mallei*. In last two decades, the occurrence of outbreaks amongst equines is steadily increasing and thus is currently considered as a re-emerging disease. In India, the disease has re-emerged in several parts of the country during recent years. Glanders is commonly manifested in three forms namely pulmonary, nasal and cutaneous glanders or Farcy. These forms are not clearly distinct in most outbreaks, and may occur simultaneously. Chronic forms are more common though, the acute form typically progresses to death within about a week. The acute form is more common in donkeys and mules than in horses. Many cases of glanders are latent and clinically asymptomatic. Respiratory tract discharges and skin are potent source of disease transmission. Animals to animals and animals to man transmission is by inhalation, ingestion of contaminated material or through skin abrasions. It is easily transmitted to humans, with 95% case fatality. All infected or potentially infected material must be handled in the laboratory under conditions of strict biocontainment. Glanders has been eradicated from many countries by statutory testing, elimination of infected animals, and import restrictions. The disease is still prevalent in Eastern European, Asian and African countries. *Burkholderia mallei* are Gram-negative nonsporulating, nonencapsulated, nonmotile rods. The bacteria grow aerobically and media containing glycerol augment their growth. Guinea-pigs are highly susceptible, and males are used for testing of potentially infected material. Intraperitoneal injections are given to attempt to elicit the Strauss reaction. The mallein test is a sensitive and specific clinical test for diagnosis of glanders, where mallein PPD is inoculated intradermopalpebrally and a delayed hypersensitivity reaction is indicative of the presence of the disease. The complement fixation test is the most accurate and reliable laboratory test available. Many other tests are now being used including enzyme-linked immunosorbent assays (ELISAs) and various PCRs. The close genetic relationship between *B. mallei* and *B. pseudomallei*, together with the high genetic variability of *B. pseudomallei*, complicates the differentiation of these organisms by either serological or molecular tools. No suitable vaccine is available at present. Equines play a pivotal role in man?s life since antiquity. This relationship still exists which is evident from a wide variety of roles equines play in the present mechanized world, including agriculture. In India, it is more significant in mountaineous and arid terrains, where approach through motorable roads is not possible. Further, army is maintaining equines for transport of logistics especially in difficult hilly terrain and a perceptible threat is always there for its eruption in indigenous population. Because of intricate co-existence of human and animals, zoonoses find a special position in India, which has a paradoxical situation to deal with. On one hand, age-old diseases affect indigenous equines and may act as a potential source of infection to elite horses while on the other hand, germplasm of exotic origin pose the danger of the ingress of exotic diseases to indigenous equids. This situation is not known in any developed part of the world. Many diseases have not been understood clearly making it difficult to implement effective control or preventive strategies. Thus mitigation of problems in India needs special and specific strategies. Among the most important bacterial pathogens is Glanders - a highly contagious disease primarily of equids. It is an age-old respiratory disease of equines occurring in different parts of the world. The disease is caused by the gram-negative bacillus *Burkholderia mallei* (the name changed from *Pseudomonas mallei* and was previously classified variously as *Pfeifferella, Loefflerella, Malleomyces or Actinobacillus*). The disease is of high zoonotic importance since it poses a significant human health risk. In humans, this usually results in fatal disease. The disease has sporadic prevalence limited to certain pockets in India. Any case of glanders if diagnosed in the field on clinical symptoms, it needs a timely confirmation for applying early mitigation and control strategy. Because there is no reliable treatment and vaccine available for the disease, it is important to have a quick and effective diagnostic tool with more sensitive, specific and less time consuming assay system. Though primarily a disease of equids (donkeys, mules and horses in order of susceptibility), carnivores are susceptible to it. Among equids, donkeys exhibit an acute form while horses tend to experience the disease in more chronic form. Mules are intermediate in susceptibility to the disease. Felids tend to be more susceptible than canids. Chronic and latent infections can occur in horses and mules. Susceptibility of laboratory animals like guinea pigs and hamsters to infection makes them good models in diagnosis of the disease. Humans (lab workers, vets, ferriers, animal attendants, sais *etc.*) get exposed to the disease generally through occupational practices. Though goats and camels can be infected, cattle and swine are resistant to the infection. Considering that the status of the disease in India, it is crucial to get the diagnosis quickly, in the field based on clinical symptoms (tentative) and timely confirmation in the lab, for applying early mitigation and control strategy. Hence we need to be prepared for its diagnosis with more sensitive, specific and less time-consuming assay system.

Current Status : Glanders was widespread throughout the world, however, it has been eradicated from many countries by employing judicious and stringent test and elimination programs. Presently the disease has not been there in European, American and Australian countries. The disease is still prevalent in Asian, African, Mexico and South American countries. In last two decades, the occurrence of outbreaks amongst equines is steadily increasing and thus is currently considered as a re-emerging disease.

Status of glanders in India

Global status

The earliest description about the disease in donkeys (384-322 BC) indicates the initiation of the disease from the neck, the head, and oozing of a thick nasal discharge from the nostrils. Later the disease was described in 4th, 5th, 14th and 17th Centuries. Major investigation is available in 18th Century when the pulmonary and cutaneous forms of the disease were described and 19th Century when causal agent was identified, mallein was prepared and used for testing of the disease. Glanders is reported from Middle Eastern countries, the Indian subcontinent, Southeast Asia, parts of China and Mongolia, and Africa. Sporadic cases are also reported from South America including Mexico and Costa Rica. The disease has been reported in Turkey, Syria, Kuwait, Iran, Iraq, Afganishtan, Pakistan, Nepal, Mongolia, Italy, India, China, Burma, Indonesia, Phillipines, Russia, Egypt and certain parts of Africa. On the other hand, glanders has been eradicated from certain parts of the world through test and elimination policy. The disease has been completely eradicated from the United States, England, Canada and Australia.

India: Glanders has been a major concern in India. Cases of glanders have earlier been reported from different parts in post independence era, most of them from north India. Several cases of glanders were recorded between 1881 and 1884 on Bareilly-Ranibagh Dakline and Bombay and Nuserabad. Glanders has been reported among civilian mules in Hisar (1962) and from mules of army (1964). A detailed account of incidence and epidemiology of glanders on military farms in India had been reported in 1975 and 1981 indicating high morbidity in the areas where horses, donkeys, and mules were thickly populated. Sporadic incidences and outbreaks in certain pockets of India have also been reported in 1984, 1985, 1988, 1990 and 1991. A case of cutaneous glanders in 1998 in a mule from Rohtak, Haryana was published which was the last confirmed case before the current outbreak in India.

In India, the disease is reported to be limited to certain pockets. The disease was not there in India from a long time but re-emergence in several parts of the country since 2006 as a major outbreak has been reported. In the outbreak during 2006-2007, several clinical cases showing typical nasal/cutaneous form of glanders were found in Maharashtra, Punjab, Uttar Pradesh, Uttrakhand, Andhra Pradesh, Himachal Pradesh and Haryana. After a gap of two years, more cases have been detected in Chhattisgarh and Himachal Pradesh in 2009-2010. Major clinical symptoms include nasal ulcers, nodular swellings all over the body, edematous swelling or wounds on legs, cough, dyspnoea, reduced appetite and wasting. Three forms *i.e.* cutaneous, nasal and pulmonary forms of glanders described in the literature were not distinct in different animals as reported earlier by many workers. Since the disease appears to have started subsequent to procurement and movement of equines from northern parts of the country, it is likely that the nidus of infection may be present in North India.

On confirmation employing field diagnosis by history, clinical signs and mallein IDP test using mallein PPD as well as laboratory assays including microbiological, serological (CFT & ELISA) and molecular tests, 160 equines from Maharashtra (26), Uttar Pradesh (70), Uttrakhand (21), Punjab (3), Himachal Pradesh (6 & 4), Andhra Pradesh (16), Haryana (1) and Chhattisgarh (13) were declared positive and eliminated. The CFT titres of serum antibodies in the positive animals ranged from 1:16 to 1:256. No human case has been reported from the area with any symptom simulating glanders. Concerned authorities of State Animal Husbandry departments and Department of Animal Husbandry, Dairying and Fisheries (Ministry of Agriculture, Govt. of India) were appropriately notified. The veterinarians and animal owners in different areas were given guidelines on sanitary measures and hygienic precautions. Strategies for prevention of further spread of the disease to other areas and neighbouring States as well action as per Glanders and Farcy Act (1899) were also suggested. Intense follow-up surveillance programme has been instituted by the Centre after these outbreaks wherein more than 15000 samples have been tested from different states of the country in last 3 years. The Centre is actively monitoring the situation nation-wide. This was important in view of the possibility of latent or clinically inapparent cases, so systematic testing is essential to identify all infected animals in an outbreak.

Transmission: The equine population contracts the disease from affected or latently infected animals. Secretions of affected animals harbour heavy loads of organism and ingestion of organism is the major route of transmission. Though inhalation can also be the route, it is less likely to cause typical disease. Direct contact through abraded skin can be minor route. Transmission is enhanced by shared food and watering facilities. Carnivores acquire infection by consuming glandered meat. In humans, however, direct contact with infected animals causing the organism to get transmitted through abraded skin and mucous membranes is considered the major route. Inhalation and ingestion can also cause the disease to be transmitted to humans. Though person-to-person transmission is rare, laboratory-acquired glanders has been reported.

Clinical Signs: After natural infection, it may take few days to several months before the manifestations are apparent. The incubation period for cutaneous form may vary from 1-5 days while that of pulmonary or septicaemic form can range from 1-2 weeks. Chronic forms and latent infections are features of glanders pausing a challenge to the veterinarians in diagnosis and elimination of cases. The disease is commonly classified into three types:

- *Pulmonary glanders*: Characterized by formation of round, greyish, firm, encapsulated nodules embedded throughout the lung tissue. The nodules contain yellowish, cheesy pus.
- *Nasal glanders*: Typified by formation of nodules in mucous membrane of nasal cavity, particularly of septum. Many of these nodules rupture and liberate mucopurulent exudates which become mixed with the serous or mucopurulent exudates already discharging from the nostrils. This yellowish green exudate is highly infectious. The ruptured nodules form ulcers with irregular, raised and hyperemic borders. The ulcers heal slowly to form a stellate scar. Regional lymph nodes are commonly swollen.
- *Cutaneous glanders or Farcy*: The typical glanders nodules form along the lymph vessels between affected lymph nodes. These nodules often rupture through skin to discharge yellowish pus and form deep ulcers, which heal slowly.
- In most outbreaks, these forms are not clearly distinct and may occur simultaneously in an animal. Chronic forms are more common though, the acute form typically progresses to death within about a week. Morbidity in donkeys, mules and horses is high especially in concentrated populations with a high rate of mortality (upto 90% in untreated cases). Sometimes donkeys even do not exhibit any apparent signs of infection before the death (especially in acute pulmonary form).

Diagnosis

- The field diagnosis is based on the history of animal movement/ new entry in the population, epidemiological pattern, clinical signs and symptoms. Typical nodules, ulcers, scars, and a debilitated condition can be sufficient to diagnose glanders. The mallein test has been the mainstay of field diagnosis. Inoculation with mallein may trigger a humoral serologic reaction to the complement fixation test.
- This seroconversion is thought to be transient but may be permanent if the animal undergoes repeated mallein testing. This is extremely important to consider if samples from the animals are destined for complement fixation test. Further, advanced clinical cases in horses and acute cases in donkeys and mules

may give inconclusive results requiring additional methods of diagnosis to be employed.

- In the lab, a number of methods ranging from isolation and identification of the agent, laboratory animal inoculation, serological methods (Complement Fixation Test (CFT) being the most widely used) and polymerase chain reaction (PCR) can be used. For eradication of glanders, CFT has been found as the preferred diagnostic tool because of its capacity to detect clinically inapparent carriers and chronically infected horses.

A. Field Diagnosis

1. Clinical Signs and symptoms

- Glanders is a contagious and fatal disease of horses, donkeys, and mules. The disease causes nodules and ulcerations in the upper respiratory tract and lungs. Typical nodules, ulcers, scars, and a debilitated condition can be sufficient to diagnose glanders. Unfortunately, many cases of glanders are latent and clinically inapparent. Therefore, systematic testing is essential to identify all infected animals in an outbreak.

2. The mallein test (a prescribed test for international trade)

- The mallein test has been the main stay of field diagnosis. Mallein PPD is a lysate (solution of water-soluble protein fractions of heat-treated *B. mallei)* containing both endotoxins and exotoxins elaborated by the organism. Infected animals are allergic/ hypersensitive to mallein and exhibit local and systemic hypersensitivity after mallein inoculation similar to that exhibited in tuberculin testing. Inoculation with mallein may trigger a humoral serologic reaction to the complement fixation test.

- This seroconversion is thought to be transient but may be permanent if the animal undergoes repeated mallein testing. This is extremely important to consider if samples from the animals are destined for complement fixation test. Further, advanced clinical cases in horses and acute cases in donkeys and mules may give inconclusive results requiring additional methods of diagnosis to be employed.

Methods

The intradermo-palpebral test

- This is the most sensitive, reliable and specific test. The preferred method of application of mallein is intradermo-palpebral. The concentrated mallein PPD (0.1 ml) is injected intradermally into the lower eyelid. The test is usually read at 24 and 48 hours after injection.

- A positive reaction is characterized by marked oedematous swelling of the eyelid, and there may be a purulent discharge from the inner canthus or conjunctiva accompanied with photophobia, pain, and depression within 12 to 72 hours. This is usually accompanied by a rise in temperature. With a negative response, there is usually no reaction or only a little swelling of the lower lid.

The ophthalmic test

- This is less reliable than the intradermo-palpebral test.

The subcutaneous test

- This test interferes with subsequent serological diagnosis. Dilute mallein is injected subcutaneously into the skin on the neck.

B. Laboratory Diagnosis

- A whole or section of a lesion, swabs of the exudates and a serum sample may be collected aseptically. The samples are required to be shipped to laboratory under refrigeration as soon as possible. Air-dried smears of exudate on glass slides may also be submitted for microscopic examination.

1. Isolation and identification of the agent

- The causative organism may be cultured from samples from fresh lesions or lymph nodes. It may also be demonstrated microscopically in films made from this material. The organisms are fairly straight Gram-negative rods with rounded ends, 2-5 ìm long and 0.3-0.8 ìm wide with granular inclusions of various size, often stain irregularly and do not have capsules or form spores, are non-motile. The organism is aerobic, growing optimally at 37°C. It grows well, but slowly, on ordinary culture media, glycerol enrichment is particularly useful.

- After a few days on glycerol agar, there is a confluent, slightly cream-coloured growth that is smooth, moist, and viscid. With continued incubation, the growth thickens and becomes dark brown and tough.

2. Laboratory animal inoculation

- Guinea-pigs, hamsters and cats have been used for diagnosis when necessary. The Strauss reaction is observed when infectious material from glanders patients is injected intraperitoneally into male guinea pigs. In positive cases, the guinea pig develops localized peritonitis involving the scrotal sac. Glanderous orchitis follows with painful enlargement of the testes. Bacteriological examination of infected testes should confirm the specificity of the response obtained.

3. Serological Tests

- A variety of serologic tests for glanders have been developed. They are superior to mallein testing in sensitivity and specificity.

Complement fixation test (a prescribed test for international trade)

- The complement fixation test is widely used and acceptable test for international trade. Although not as sensitive as the mallein test, the CF test is an accurate serological test that has been used for glanders diagnosis for many years. It is reported to be about 95% accurate, serum being positive within 1 week of infection and remaining positive in the case of exacerbation of the chronic process. CFT was the most extensively used test for the diagnosis of glanders and reported to be highly sensitive by most workers.

Enzyme-linked immunosorbent assays and other tests

- Dot enzyme-linked immunosorbent assay (Avidin-biotin dot ELISA).
- Plate and membrane (blot) enzyme-linked immunosorbent assays (ELISAs).
- Dipstick dot-blot and electrophoretically separated and transferred western blot methods.
- Competitive ELISA.
- Counter-immunoelectrophoresis test.

- Rose bengal plate agglutination test (RBT).
- Polymerase chain reaction (PCR); *B. mallei*-specific real-time PCR assay targeting the *B. mallei bimA*ma gene (*Burkholderia* intracellular motility A; BMAA0749).

Continuing development of monoclonal antibody reagents specific for *B. mallei* antigenic components offers the potential for more specific ELISAs in the foreseeable future. At this time, none of these tests has been validated. However, PCR has the potential to be a safe, fast method to confirm infection. Methods to differentiate between *B.mallei* and *B. pseudomallei* are important as this cannot be reliably accomplished by the current serological tests.

Differential Diagnosis : Cases for specific glanders investigation should be differentiated on clinical grounds from other chronic infections of the nasal mucosae or sinuses, and from strangles (*Streptococcus equi* infection), ulcerative lymphangitis (*Corynebacterium pseudotuberculosis*), pseudotuberculosis (*Yersinia pseudotuberculosis*), sporotrichosis (*Sporotrichium* spp.) and other forms of pneumonia. Glanders should be excluded positively from suspected cases of epizootic lymphangitis (caused by *Histoplasma farciminosum*), with which it has many clinical similarities. In humans in particular, glanders should be distinguished from melioidosis (*B. pseudomallei* infection). Purulent sinusitis, guttural pouch empyema, and other causes of nasal catarrh should also be considered. Skin lesions may be similar to those of dermatophilosis or dermatomycoses such as sporotrichosis. Knowledge of the progressive debilitating nature of glanders and application of serological or mallein tests will serve to distinguish glanders from other similar diseases. The close genetic relationship between *B. mallei* and *B. pseudomallei*, together with the high genetic variability of *B. pseudomallei*, complicates the differentiation of these organisms by either serological or molecular tools.

Treatment and Control

Formerly the disease is notifiable in India and had to be dealt within the purview of the "Glanders and Facry Act, 1899" whereby the cases should not be treated and control is done by test and elimination policy. The Prevention and Control of Infectious and Contagious Diseases in Animals Act 2009 has been *in vogue* as a Central Act to deal with this dreaded infection. The organism is sensitive to many antimicrobials *in vitro*, but the risk of transmission and spread to other susceptible populations warrants that affected animals be eliminated. In humans, many antimicrobials have been used including sulphonamides for treatment. No suitable vaccine is available at present. Control measures for successful containment of the disease include surveillance and monitoring of the disease in target populations, restriction on movement of animals, elimination of affected cases, proper disposal of carcasses, disinfection of premises, other zoo-sanitary and personal hygiene measures. The organism is relatively sensitive to common disinfectants and desiccation and can be readily killed by direct sunlight. However, the organism is reported to survive for upto 6 weeks in infected stables.

Perspective: Considering the need for quicker and reliable diagnosis with more sensitive, specific and less time-consuming assay system, NRCE is working on the development of some diagnostics including indirect ELISA, dot-ELISA and Indirect Haemagglutination (IHA) test *vis a vis* Complement Fixation Test using conventional and recombinant antigens. One such recombinant protein expressed from synthetic gene by DRDE has been proved to be useful in ELISA by NRCE. NRCE is studying molecular epidemiology of the disease and development of quicker, specific and sensitive diagnostics using synthetic peptide or recombinant proteins which would help in eradication of the disease from the country. The Centre is also concentrating on development of other rapid and specific tests with higher sensitivity employing molecular tools viz. PCR. This is more important in view of the "Infectious Disease Act 2009" *in vogue* which has the provision of elimination of positive reactors. Thus screening of more samples would help pointing to the nidus of infection and their elimination. The eradication of the disease from the country with the help of strong coordination and network of animal husbandry authorities in different states is quite achievable.

Avian Influenza (Bird Flu)

Being an acute infection with very short incubation period sudden death of chickens in large numbers (nearing to 100%) without clinical signs is the most prominent feature in HPAI (H5N1) outbreaks. Similar clinical signs and mortality pattern is also observed in Newcastle disease (Ranikhet disease) affecting poultry which are unvaccinated. The presence of proteolytic bacteria in the lower digestive tract of poultry suggests the possibility in cleavage of hemagglutinin that may alter the pathogenicity of avian influenza viruses. Swelling of head, nasal discharge, greenish diarrhea, swelling and redness of comb, wattle and foot pad and hemorrhages in internal organs and skeletal muscles are prominent pathological changes. Because of the extensive hemorrhagic changes the disease in chicken is also known as "**Chicken Ebola**". Pathognomonic pulmonary changes include extensive hemorrhages and accumulation of edematous fluid in alveoli and infiltration of mononuclear cells. Other changes include intracranial hemorrhages, congestion of meninges and accumulation of glial cells in the brain. Wild water fowls including ducks do not show clinical signs and pathological changes, hence considered carriers. However in mid 2005 as many as 6000 migratory birds died in Qinghai Lake which is unprecedented. In chickens antibody rarely develop as they die of acute infection where as antibody is produced in humans. Pathological changes in humans are more or less similar to that of avians. The first signs of the disease develop 2 to 4 days after the last exposure to infected sick poultry, but longer incubation period of up to 8 days have been reported. Most patients present symptoms of fever, cough, and shortness of breath and radiological evidence of pneumonia .In the early stage main pathological findings are pulmonary hemorrhage, edema and consolidation. The alveolar and bronchiolar laminas are filled with protein rich fluid, erythrocytes, fibrin, and leucocytes. Subsequently diffuse alveolar damage, hyaline membrane formation, proliferation of pneumocytes and bronchial epithelium, fibrosis of interstitial tissues and alveolar spaces are seen . The most important feature of immunopathogenesis is the appearance of hypercytokinemia (cytokine storm) characterized by extreme production and secretion of pro-inflammatory cytokines and attributed to severity of the disease. Because of systemic infection viral antigen is detected in lungs, upper respiratory tract, small and large intestinal epithelial cells, hematopoietic cells in the bone marrow, glial cells and neurons of the brain, and lymphocytes.

Diagnosis of avian influenza H5N1 is a challenging task. Quick and reliable diagnosis is a prerequisite for successful control of the disease in poultry and treatment in humans. Conventional diagnostic tests like isolation of the virus in chicken embryos and cell cultures (MDCK cells) and identification by hemagglutination (HA) and hemagglutination inhibition (HI) tests and micro neutralization assays though considered gold standards usually take two to three weeks. Successful isolation of the virus depends on appropriate methods of sample collection, preservation and transport to the diagnostic laboratory. This hampers in initiating control measures in a reasonable time and allows the disease to spread. Laboratory confirmation with in two to three days can be made by other diagnostic tests like direct or indirect immunoflurescence tests, reverse transcriptase polymerase chain reaction (RT-PCR) and real time RT-PCR (rRT-PCR).These diagnostic tests are similar both for humans and non-humans, except for HI test in which horse erythrocytes give better results for human samples than chicken erythrocytes. For diagnostic purpose RT-PCR and rRT-PCR are handier as compared to isolation of the virus, the later being more reliable. Multiplex RT-PCR and immuno-PCR assays for rapid detection and sub- typing from clinical samples are now available. Polymerase chain reaction (PCR) has become the method of choice for virus subtype identification. However, designing subtype-specific PCR primer pairs is a very challenging task. Primer pairs must amplify heterogenic subtype sequences and also discriminate between subtype of interest and closely related subtypes. These authors developed a tool, known as **Primer Hunter**, which helps in designing the primer pair that efficiently amplifies any one of target sequences and none of the nontarget sequences. Rapid molecular sub typing and pathotyping of avian influenza viruses using SYBR Green RT-PCR assay has been reported. Sequencing of the viral DNA is important for genotyping and phylogenetic analysis, and to trace the origin of the virus. Sequencing is also used to assess the pathogenicity in chickens, adaptability to mammals (627K in PB2 gene), and oseltamivir resistance (His 274 Tyr in NA gene) etc. Rapid diagnostic kits though used for preliminary diagnosis lack acceptable level of sensitivity and specificity.

Control of avian influenza (AI) (H5N1) outbreaks is essential to safeguard the poultry population which is the major source of income of farmers and source of animal protein for human being. Its control is also essential to prevent the risk of human infection. Confirmation of infection or an outbreak of the disease in chickens calls for immediate control measures as per the WHO and OIE guidelines. By now each country, also India has developed action plan for control of the disease. Culling of chickens infected with the virus and at risk within the specified zone of 3 km is the recommended practice for control of the disease. The need for identifying a surveillance zone up to 5 km, surrounding the culling zone, is equally important for successful control of outbreaks. The culling is followed by sanitization of potentially contaminated materials and premises. Restocking is recommended three months after sanitization. Strict measures of culling, sanitization and restocking adopted by India during 2006 to 2010 outbreaks have been a success story for the control of HPAI. During this period as many as 6.63 million of chickens were culled. Large number of eggs and several tons of chicken feed were also destroyed. The poultry farmers were adequately compensated as per the criteria fixed by the government. The epidemiological and evolutionary repercussions of mass culling.was explored and it was concluded that mass culling achieves a short term epidemiological benefit at the expense of long term detriments: a more genetically susceptible host population, ultimately greater mortality and elevated influenza virulence. The other option to control the disease is by vaccination of poultry with killed heterologous (H5N2, H5N9) or homologous strains (H5N1) and other available vaccines generated through molecular techniques using fowl pox and Newcastle disease viruses as vectors .Vaccination against AI (H5N1) may result in issues related to surveillance programme and international trade of poultry and poultry products. The AI vaccination in poultry is acceptable if a reliable test is in place to differentiate infected from vaccinated animals (DIVA). The strategies to accomplice DIVA are use of sentinel birds in vaccinated flock / area, use of vaccine virus with same HA and different NA and use of nonstructural protein (NS1) as diagnostic reagent. In spite of having all these strategies an exit policy must be inplace. In developing countries vaccination programmes in avian species have been recommended. However it requires concurrent management of husbandry practices and industry compliance to eradicate the disease rather than establishment of an endemic situation arising from introduction. Potent AI vaccines, when properly used, can prevent disease and death, increase resistance to infection, reduce field virus replication and shedding, and reduce virus transmission, but don?t provide "sterilizing immunity".

Different strategies for control of AI have been suggested such as routine vaccination in endemic areas, emergency vaccination at the face of an epidemic and preventive vaccination whenever a high risk of virus incursion is identified. Many countries have adopted both culling and vaccination strategies for effective control the disease rather than culling alone and vaccination. With these measures outbreaks have been controlled to a large extent and now only 3 of the 65 countries have H5N1 virus infections in poultry. Though India has not resorted to vaccination a killed vaccine using H5N1 virus isolated from Navapur outbreak in 2006 has been developed with high level of efficacy for use in emergency. Due to acuteness of the disease, birds usually die within 48 to 72 hours of infection, treatment on individual basis is not recommended. For treatment in humans infected with H5N1 virus some antiviral drugs, notably oseltamivir is widely used. Use of this neuraminidase inhibitor has resulted in emergence of resistant strain. Recently influenza A nucleoprotein has been identified as a druggable target and found a small molecule, nucleozin, which is very effective for H5N1 virus. The pandemic threat of H5N1 has not receded in spite of taking steps to control the disease in poultry and other avian species. Fearing the unset of a pandemic due to AI (H5N1) virus or any reassortanat strain the WHO recommended the use of some strains, including one from India, for vaccine production.

Rabies

The first serious effort to control rabies would have to focus on the major transmitter of rabies i.e. dog. The total dog population is approximately 25 million. The Indian dogs fall into 4 broad categories viz. pets (restricted and supervised); family dogs (partially restricted, wholly dependent); community dogs (unrestricted, partially dependent); and feral dogs (unrestricted, independent). Most dogs in India, perhaps 80%, would fall into the last 3 categories. Until 1998, the population of stray dogs in India was kept under check by civic authorities, by impounding and euthanizing unclaimed dogs However, because of pressure from animal welfare activists, this approach was replaced by a policy of animal birth control, also referred to as the ABC Program. In this program, stray dogs are impounded, surgically sterilized and released back into the area from where they were picked up. The success of this program hinges on the sterilization of 70% of the strays in a given geographic area within 6 months, before the next reproductive cycle, otherwise the entire effort is negated. This target is difficult to achieve, unless there is a coordinated approach among various governmental as well as non-governmental organizations given the large number of strays and the limited resources. With the limited resources and the efforts, it would be a lot more productive if the focus of the sterilization is only the bitches since if all the bitches of the area are sterilized, the fertile dogs cannot reproduce whereas any left out dog would successfully reproduce with even a single bitch negating the entire exercise of population control. However, as of now, the unabated reproduction among the community and feral dogs is leading to a steady increase in the dog population and consequent cases of dog bites and increasing demand for post-exposure vaccination leaving behind any prevention and control efforts. If the decision makers continue to look the other way, India would find itself in the midst of an epidemic of rabies in dogs with heightened incidental and deliberate exposures to humans by the broad day light perpetuator of rabies viz., dog.

Latest Ante Mortem Approaches in Diagnosis of Rabies

Urgency in getting the diagnosis

- So far recommended techniques are based on brain sample examination.
- Thus, standard protocol needs to wait for the death of the suspected animal.
- There is the urgency in getting the diagnosis of Rabies especially in case of bite/lick exposure in live animals.
- So far, no standard protocol for diagnosis of Rabies in live animals.

Ante-Mortem Diagnostic Approaches

- Immunological detection of anti-Rabies Antibodies.
- Molecular detection of Rabies RNA.

Immunological detection of Rabies Antigen

- Direct Immunofluorescence test is the gold standard for detection of rabies antigen.
- Especial care should be exercised in standardizing the conjugate.
- Commercial anti-nucleocapsid FITC conjugate should be preferred.
- Direct FAT is the fast and reliable approach – a must for all authentic Rabies diagnostic laboratories.

Molecular detection of Rabies RNA by RT-PCR

- Reverse transcriptase Polymerase Chain Reaction (RT-PCR) and Real time PCR can be used for the detection of rabies.
- Smith and coworkers standardized hemi-nested RT-PCR for detection of Rabies from saliva and skin specimens.

- Real-time PCR assay has been found to be more sensitive than conventional RT-PCR assay.

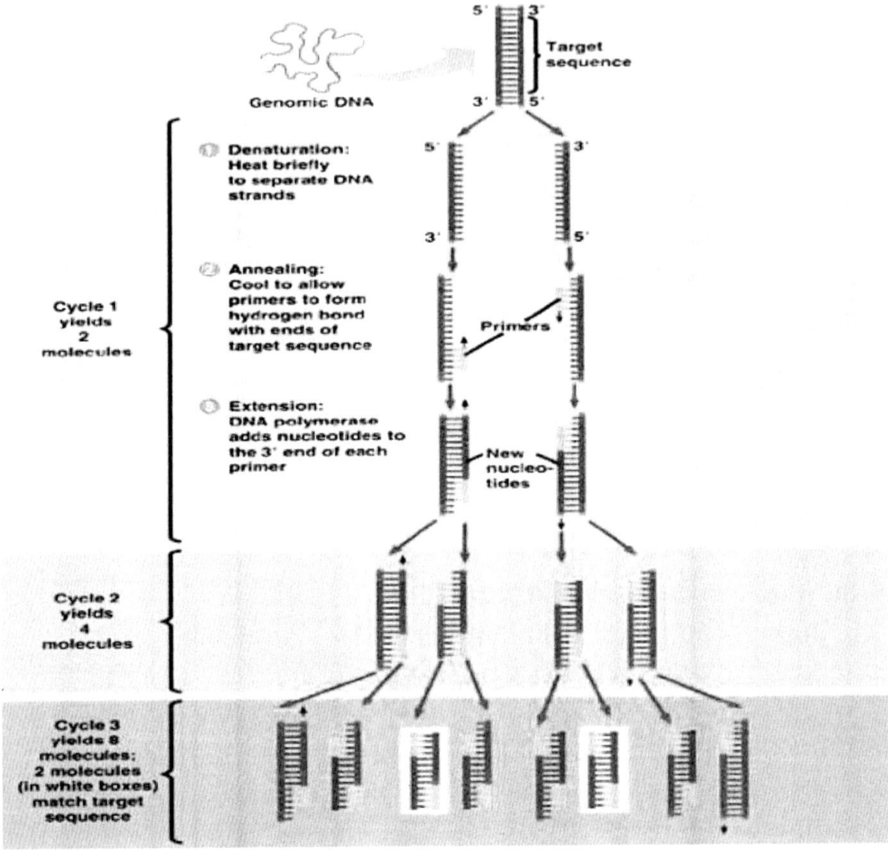

NRT-PCR, hNRT-PCR

- By using Nested or hnRT- PCR a 10- fold increase in sensitivity compared with that of the external RT-PCR can be achieved.
- Sensitivity of Nested and hnRT- PCR makes it the approach of choice for detecting limited amounts of virus in the saliva that is loaded with proteases and nucleases.
- Thus, Nested or hnRT- PCR is highly sensitive technique for detecting rabies RNA sequences in saliva.

Real-time PCR

- **Real-time PCR**, also known as kinetic PCR, quantitative PCR, qPCR, qRT-PCR or RT-qPCR.

- Real-time PCR requires a special thermocycler equipped with a sensitive camera that monitors the fluorescence in each well of the 96-well plate at frequent intervals during the PCR Reaction.
- **Intercalator-based method**, also known as **SYBR Green method**, uses a double-stranded DNA dye in the PCR reaction which binds only to double-stranded DNA and gives fluorescence.
- **Probe-based real-time PCR**, also known as **TaqMan PCR**, It is designed to bind only the DNA sequence between the two PCR primers. Therefore, only specific PCR product can generate fluorescence signal.
- TaqMan method is more accurate and reliable than SYBR green method, but also more expensive.

Immunological detection of anti-Rabies Antibodies

- The appearance of high titer of antibodies to rabies viral antigen is diagnostic in an animal with no history of pre-immunization against rabies.
- Rabies specific antibodies usually appear after 7 days of the illness in an unvaccinated patient and may be a good diagnostic marker in cases with paralytic rabies that survives longer.
- Mice Neutralization Test developed by Webster and Atanasiu in 1935. Mice neutralization tests are based on the fact that antibodies in the serum of immunized (infected) animal can lower the infectivity of a known dose of rabies virus in mice.
- Rapid Fluorescent Focus Inhibition Test (RFFIT). It precisely is a combination of cell culture and Immunofluorescence. It is an advanced approach using international standards and provides IU of quantitative antibody level.
- Fluorescent antibody virus neutralization (FAVN) - an OIE-prescribed test – can also be used.
- When there is low titer of antibody, in early phase of disease, distinguishing negative sera from the positive is easier with the FAVN.
- To differentiate infection and traces of old immunity, antibodies should also be tested in CSF and antibody titer be repeated in serum and CSF after 7-10 days to find significant increase in antibody titer.

- Sera can also be analyzed by indirect enzyme-linked immunosorbent assay (ELISA) wherein glycoprotein is used as antigen.
- Preferably, a kit for the purpose (Platelia Kit, Bio-Rad Laboratories, (France) may be used.
- Some workers reported Counter Immuno Electrophoresis (CIE) for field detection of rabies antigen and modified CIE for quantitative analysis of anti-rabies antibodies.

Serology Techniques – MNT

- Webster and Atanasiu developed Mice Neutralization Test (MNT) in 1935 which is a widely used WHO recommended test and has become the standard against which other tests are now evaluated.
- To each serum dilution from a suspected case of rabies, virus is added and the mixture is incubated for 90 min at 37°C and then inoculated intra-cerebrally into weanling mice (5-10 mice for each serum dilution).
- The number of deaths occurring in each serum dilution is noted.
- The dilution of serum at which 50% of the mice survive is calculated by methods like Reed & Muench or Spearman & Karber.
- The precision of the assay is limited by the number of mice inoculated for each serum dilution and the increment between dilutions to be used. Tediousness and economics limit the numbers of mice, to be inoculated and dilutions to be evaluated.
- The individual resistance and nonspecific mortality associated with any animal test also adversely affect the precision of the test.

Serology Techniques – RFFIT

- The RFFIT is one of the most widely used substitutes to the Mouse Neutralization Test (MNT).
- It is a rapid test that requires only 20 h for completion.
- It involves the observation of foci of virus-infected cells by fluorescent antibody staining.

- The sera are serially diluted in 8-well Lab-Tek slides, followed by the addition of predetermined FFD 50 (Focus Forming Dose) of virus [One FFD 50 is the dilution of virus at which 50% of the observed fields contain one or more infected cells].
- The serum-virus mixture is incubated at 37°C.
- Subsequently, suspension of cell line is added.
- After a 20-h incubation period at 37°C in a 5% CO2 incubator, the cells are stained with an anti-rabies antibody conjugated to FITC.
- The antibody titer is defined as the reciprocal of the serum dilution, which reduces the challenge virus to 1 FFD 50.
- The titers obtained are approximately equal to those measured by the MNT.
- The rapidity with which the test can be carried out and the ease with which several samples may be tested at a time make it suitable for routine diagnosis.
- The RFFIT is more sensitive than the MNT for detecting Virus Neutralizing antibodies (VNAs).

Serology Techniques – FAVN

- It is an adaptation of RFFIT, was developed by Cliquet and coworkers in 1998.

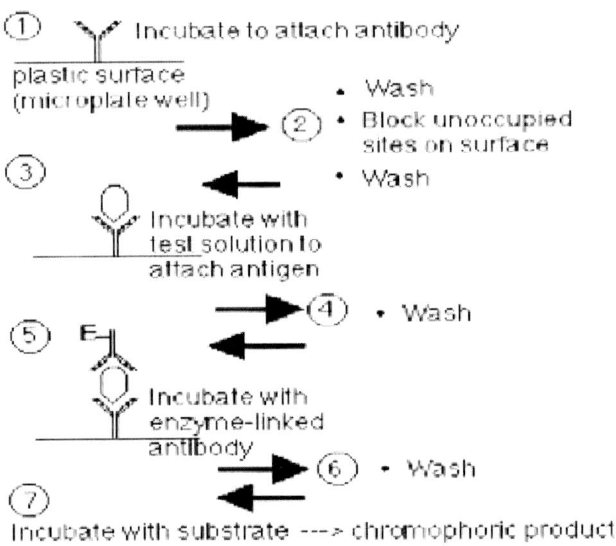

- The diluted serum is placed in four wells of a microtiter plate containing virus in cell line and each well is scored as having virus present or no virus after a 40-h incubation period.
- Results obtained with this test showed good agreement with the MNT and RFFIT.
- Hostnik has described a modification of this test in which the monoclonal anti-rabies antibodies and a peroxidase anti-mouse conjugate were used instead of a fluorescein-conjugated anti-rabies antibody.

Serology Techniques – IFA

- The bound antibody is visualized using a fluorescein-labeled species-specific antiserum.
- The slide is examined under a fluorescent microscope.
- The reaction is graded by the amount and brilliance of fluorescence.
- The IFA is reported to be more sensitive than the mice neutralization test (MNT) in detecting antibody early in the immune response, making it valuable for diagnosis.
- The IFA can also be used in sero-epidemiological surveys.
- It is a rapid test and is economical and easy to perform.

Serology Techniques – CIEP

- Agarose gel (0.9%) is prepared in barbital buffer (pH 8.2). Each microscopic slide is layered with molten agarose and allowed to solidify at the room temperature.
- Various dilutions of test sera viz. 1:2, 1:4, 1:8, 1:16 are prepared in PBS (pH 7.2).

- Each dilution of test serum is mixed with a fixed amount of antigen and incubated at 37°C for one hour in water bath and are added to 6 mm wells.
- Slides are placed in electrophoresis chamber and run for 45 minutes at 10 volts/slide.
- After the run agar is removed from 3 mm wells and indicator serum is added in these wells. Thereafter, electrophoresis is continued for another 2 hours at a constant current of 15 mA/slide.
- The gels are dried at 37°C by keeping a wet filter paper over them.
- The slide is stained with one per cent Coomassie brilliant blue stain for 15 minutes and then destained with destaining solution for 2 minutes.
- The slides are observed for precipitation lines.

Detection from saliva

- As the secretion of rabies virus into the saliva is intermittent, its isolation from saliva may not always yield successful results.
- Detection of Rabies RNA.
- Detection of Rabies antigen.
- Isolation of virus in cell culture.

Detection from CSF

- Detection of anti-rabies antibodies.
- Detection of Rabies RNA.
- On rare occasions, virus has been isolated from the CSF.
- Detection of Rabies immune complexes.
- Clinico-pathological estimation of CSF reveal significant pleocytosis.

Detection from skin biopsy

- Detection of Rabies RNA.
- Detection of Rabies antigen.

- Performed on frozen sections of the skin biopsy samples, exhibited the highest sensitivity.
- The section of skin that having diameter of 5-6 mm should be taken.
- The biopsy section should involve examination of a minimum of 10 hair follicles.
- The biopsy must be of sufficient depth to include the cutaneous nerves.
- No preservatives be added.
- The tissue is fixed in acetone and subjected to FAT on cryosections.
- The antigen is usually present in the nerve fibers surrounding the base of hair follicles. Typical rabies nucleocapsid inclusions can be observed in the nerve endings around the base of some of these hair follicles.
- At least twenty serial sections be examined before giving a negative result.
- The test becomes more sensitive as the disease progresses.

Detection from urine

- Detection of Rabies RNA.

Corneal Impression Smear Examination

- Detection of Rabies antigen.
- Direct FAT test on corneal epithelium.
- The corneal smear first developed with mice by Schneider in animals and by Cifuentes and coworkers in humans.
- Reliability of corneal impressions for rabies diagnosis is limited, since a negative result can not rule out the diagnosis of rabies.

Infrared Thermography

- A noninvasive, non-contact diagnostic technique.
- It measures heat emitted from a target surface.
- Infrared radiation is detected by a thermal camera.
- It provides pictorial representation of the thermal information.

CT Scan- MRI

- Computed Tomographic (CT) studies have revealed hypodense lesions in cortex and basal ganglia.
- Magnetic Resonance Imaging (MRI) of the brain, however, increased signal intensities in grey matter areas.
- MRI has also revealed ill-defined hyperintensities in the brain stem, hippocampus, hypothalamus, subcortical and deep white matter and cortical grey matter.
- at the stage of terminal coma, contrast-enhanced lesions are observed in the brain stem, hypothalami and spinal nerve roots.

Symptoms

- In spite of all intra-species and inter-species variations and those that are observed at different phases of the disease.
- Prominent characteristics are attributed to each species.
- Peculiar symptoms during various phases of the clinical disease among animals of various species.

Conclusion

- Rabies is a neglected disease in India. The pathological alterations are characteristic though not consistent. It will be necessary to mount a vaccination program to attempt to protect 70% of dogs, a strategy that has proven effective in other countries. There is also a need for post-exposure vaccination programs for men and animals, the level to be subject to cost-benefit studies.
- The basic philosophy regulating rabies control programs in man and animals includes a systematic reduction of vector populations and enhancement of their immune status by vaccination supported by strict surveillance and reinforced regulation of quarantine of imports. Vector population density and turnover are, therefore, the key points of concern in this approach.
- The effective control of rabies among dogs in India would be possible with the proper disposal of garbage with serious participation of municipal authorities; reductions of the stray dog and stray livestock populations by legislation whereby disallowing stray animals as such; and by mass vaccination of dogs by effective usage of bait and parenteral vaccines.

Parasitic Zoonoses

Global agencies like FAO and WHO have recommended continued greater surveillance of prarasitic zoonoses in man and animals. At the Third Global Meeting for parasitic control held at WHO headquarters, Geneva in November, 2004, food-borne trematodosis including fasciolosis, were added to the list of helminthosis with a great impact on human development. The increased incidence of established protozoan zoonoses as well as growing evidence of emerging parasitisms give cause for concern to various stakeholders for public health.

- It is imperative to have current information regarding prevalence of zoonotic parasites and the risk factors associated with human infections. This will help in more effective implementation of strategic control programme. Individuals with HIV or other immuno-compromised states must be provided protection from contamination of food and water. Routine stool examination should desirably include new entities such as sporocysts of Sarcocystis and Cryptosporidium in the ambit of reportable infectious material. The role of the veterinarian is paramount in coordinating with medical and public health personnel for the common objective of minimising the challenge to health and economy posed by FBPZ.
- Climatic changes due to global warming, increased vector population, world tourism, demand for livestock food products, changing socioeconomic conditions, poverty, lack of personnel hygiene, defecating in open space, scarcity of potable water, abundance of stray animals, high population density and culinary habits are responsible for rising zoonoses in India.
- Emergence of diseases like human hydatidosis, neuro-cysticercosis, cryptosporidiosis, toxoplasmosis in acquired immune deficiency syndrome and reemergence of cutaneous leishmaniasis pose a great threat and challenge for prevention and control of parasitic zoonoses in India.
- Use of molecular epidemiological investigations, spatial analytical tools can help better understand current status of parasitic zoonoses in India.

- Health education, vector control, sanitary movements, controlled slaughtering, higher socio-economic development and improved risk assessments can help control parasitic zoonoses in the country.
- There is need to have better coordination between medical and veterinary scientists in order to formulate appropriate control strategies.

❐❐❐

Chapter 16

Strategies for Carcass Disposal During Disaster

- Thousands of animals died every year due to
  - Natural disasters: Flood, Earthquake, Drought etc.
  - Man-made disaster: Bhopal gas tragedy etc.
  - Infectious disease outbreaks : Bird flu epidemic etc.
- Many difficulties encountered by officials involved in carcass disposal management during disaster worldwide.

- There is need for the establishment of efficient and effective advance planning mechanisms, to mitigate the consequences of carcass disposal situations.

Carcass Disposal Methods

The following chapter provides a broad overview of the most commonly utilized methods for animal carcass disposal. Each of the five major methods possesses strengths and weaknesses that make them the selection of choice in certain situations, and the same will preclude their consideration in other scenarios. It is essential that emergency response officials understand the consequences of selecting (or not) specific disposal methods as part of any response procedure.

- Burial
- Incineration (Burning)
- Composting
- Rendering
- Alkaline Hydrolysis

Burial

Burial method of animal carcasses disposal is the preferred method of choice when infectious agents are involved (except where prion agents are suspected), but it can also be generally used in natural disasters. It is preferred because it is generally quicker, cheaper, environmentally cleaner, easiest to organize, and often the most convenient means of disposing of large numbers of livestock there are two common methods of burial for animal carcasses. Open pit disposal has been the most common method used by commercial poultry producers for disposing of dead animals. However, burial in disposal pits poses a threat to groundwater quality. The carcasses can leach contaminants for an undetermined length of time if they do not decompose properly. Ambient temperature and moisture conditions can slow or speed up the degradation process, thus affecting environmental contamination possibilities as well. Open pits are also susceptible to scavenger intrusions which are highly undesirable in disease related disasters. Therefore freshly closed pits have become the method of choice for the most current disaster situations. By heaping soil on top of the pit, the weight of the soil acts to stop carcasses from rising out of the pit due to gas entrapment, prevents scavengers from digging up carcasses, helps filter out odors, and assists in absorbing the fluids of decomposition. Burial site selection is a critical component of the decision process when first deciding if burial is a feasible alternative, and secondly where to dispose of the animals. An unacceptable burial site can create health, environmental, and aesthetic problems for the disaster response officials. There are many important considerations to explore in choosing a proper burial site such as:

- Access to site.
- Facilities available.
- Equipment required.
- Safety to personnel.
- Acceptability to owner of property.
- Protection from public view.
- Height of water table.
- Distance from residences/roads.
- Surface slope.
- Cultural/historical considerations.
- Distance from streams or wells.
- Bio security considerations.

- Preferred method of choice when infectious agents are involved.
- It can also be generally used in natural disasters.
- Preferred because it is quicker, cheaper, environmentally cleaner, easiest to organize, and most convenient means of disposing of large numbers of livestock.
- Burial in disposal pits poses a threat to ground water quality.

Types of burial disposal

1. Open pit disposal
2. Closed pits disposal

Open pit disposal

- Most common method used by commercial poultry producers for disposing of dead birds.
- Ambient temperature and moisture conditions can slow or speed up the degradation process.
- Also susceptible to scavenger intrusions.

Closed pits disposal

- Method of choice for the most current disaster situations.
- Heaping soil on top of the pit.
- Weight of the soil acts to stop carcasses from rising out of the pit due to gas entrapment.
- Prevents scavengers from digging up carcasses.
- Helps filter out odors.
- Assists in absorbing the fluids of decomposition.

Incineration (Burning)

Burning of animal carcasses produces a solid waste by-product (bone and ash) that is essentially free of pathogens or putrid material if done properly. For these reasons, it is a desirable form of carcass disposal in many situations. There are limiting factors, however, that can preclude its consideration. Some of the factors are:

- Location of site
- Access to site
- Type of animal carcass involved
- Fuel availability
- Amount of carcasses to burn
- Environmental considerations

There are three commonly used methods of incineration. They include open-air burning, biological incineration, and controlled burning. Open air burning of animal carcasses requires addition of combustible material such as timbers and straw as fuel additives to achieve sufficient temperatures to completely consume the carcasses. Smoke from such fires can be high in particulates and/or produce offensive odors if the burn is not complete.

The type of animal to be disposed of will also play a critical role in the success of open air burning as the method for consideration. Animals with high fat content such as hogs will burn much faster and with less fuel requirements than poultry who are low in fat, and whose feathers do not burn easily.

Biological incineration is an efficient disposal method, as it creates almost no pollution or particulates, and achieves virtually complete oxidation of the carcasses.

There are limiting factors for this method unfortunately, such as cost, lack of portability, location of existing incinerators, and capacity restraints. The third type of burning technique is controlled burning such as in an open pit, or by air curtain incineration. Air–curtain incinerators (also called Trench burners) are a relatively new technology that is now used in many large-scale natural disasters to burn combustible debris. The incinerators consist of large capacity fans driven by diesel engines connected to ducting, which delivers the high velocity air down into a long narrow pit or trench. A commonly recommended dimension for the trench is 8x8x35 inch, but the size can be altered according the amount of carcasses or debris to be consumed. The system delivers the air stream at approximately 165 miles per hour down into the pit at an angle to create a "mini-cyclone" within the pit. The continual downward pressure by the incoming air forces the complete destruction (burn) of all material with very little smoke produced, at temperatures of up to 2000° Fahrenheit. The advantages of the air curtain incinerators are that they are portable, environmentally friendly (minimal ash or particulates produced), and can incinerate vegetative debris from natural disasters (as a fuel source) at the same time the carcasses are consumed. Some disadvantages are that the incinerators are expensive to operate, are not available in all locations (no known state or federal entity owns any), and may require excessive fuel depending on the material to be incinerated.

- Burning of animal carcasses produces a solid waste by-product (bone and ash) that is essentially free of pathogens or putrid material if done properly.
- It is three type:
  1. Open-air burning
  2. Biological incineration
  3. Controlled burning

Open-air burning

- Open air burning of animal carcasses requires addition of combustible material such as timbers and straw as fuel additives to achieve sufficient temperatures to completely consume the carcasses.
- Smoke from such fires can be high in particulates and produce offensive odors if the burn is not complete.
- The type of animal to be disposed of will also play a critical role in the success of open air burning as the method for consideration.

Biological incineration

- Efficient disposal method
- Creates almost no pollution or particulates.
- Achieves complete oxidation of the carcasses.
- Limiting factors for this method unfortunately, such as cost, lack of portability, location of existing incinerators and capacity restraints.

Controlled burning

- Open pit incineration
- Air curtain incineration or Trench burners.

Air Curtain Incineration or Trench burners

- New technology that is now used in many large - scale natural disasters to burn combustible debris.
- The incinerators consist of large capacity fans driven by diesel engines connected to ducting, which delivers the high velocity air down in to along narrow pit or trench.
- Advantages of the air curtain incinerators are that they are portable, environmentally friendly (minimal ash or particulates produced), and can incinerate vegetative debris from natural disasters (as a fuel source).

Composting

Composting is defined as the controlled decomposition of organic materials. Decomposition occurs when organic materials go through a "slow cooking" process as heat and microorganisms consume the organics. Composting consists of two stages, a primary high temperature active stage, and a secondary lower-temperature "curing" or stabilization stage. The primary phase of composting takes 2-3 months and the secondary phase another 2-3 months. The end result of the process is the production of carbon dioxide, water vapor, heat and compost.

Compost is considered to be one of the more environmentally friendly forms of carcass disposal, because it is in effect a form of recycling. It is applicable for many natural disaster situations and is routinely used in the commercial poultry industry today as an accepted form of disposal. It can be applied to large animals in some cases, especially swine, but is not appropriate when disease biosecurity is an issue. In many states though, composting is not legal for large animal carcasses because of the time (3-6 months) it takes to complete the process. The advantages of composting are that initial start up costs are minimal, and the end product can be utilized as fertilizer material or a soil additive. Composting disadvantages are that it is a slow process (months) which requires some monitoring throughout the process, and is usually not appropriate for disease situations because the causative organisms may not be destroyed immediately.

- Controlled decomposition of organic materials.
- Decomposition occurs when organic materials go through a "slowcooking" process as heat and microorganisms consume the organics.

- Composting consists of two stages, a primary high temperature active stage, and a secondary lower-temperature "curing" or stabilization stage.
- The primary phase of composting takes 2-3 months and these condary phase another 2-3months.
- The end result of the process is the production of carbon dioxide, water vapor, heat and compost.

- Considered to be one of the more environmentally friendly forms of carcass disposal, because it is in effect a form of recycling.
- Applicable form any natural disaster situations and is routinely used in the commercial poultry industry today as an accepted form of disposal.
- It can be applied to large animals in some cases, especially swine, but is not appropriate when disease biosecurity is an issue.
- The advantages of composting are that initial start up costs are minimal, and the end product can be utilized as fertilizer material or a soil additive.
- Composting disadvantages are that it is a slow process (months) which requires some monitoring throughout the process, and is usually not appropriate for disease situations because the causative organisms may not be destroyed immediately.

Rendering

Rendering is a process of separating animal fats and proteins, usually by cooking. The recovered proteins are used almost exclusively as animal foodstuffs, while there covered fats are used both industrially and in animal feeds. There are two primary methods of rendering. The older method uses steam under pressure (with a grinding process) in large closed tanks. A second and newer method is dry rendering, which cooks the material in its own fat by dry heat in open steam-jacketed drums.

Rendering is considered an environmentally friendly method of disposal because it recycles the animal protein from the carcasses back into a usable form as meat or bone meal. Nevertheless it has limitations.

Rendering is not economically feasible for poultry, and most renderers have voluntarily declined to render sheep carcasses because of concerns that the rendering of sheep infected with Scrapie (which were then used as protein supplements in cattle feed).

The environmentally friendly concept, along with the production of a marketable product is the main advantages of rendering. Disadvantages include the fact that rendering is not always appropriate for disease situations because the carcasses must be transported to the plant, some species are not amendable to efficient rendering practices, and in many areas there is a lack of available rendering facilities.

- Process of separating animal fats and proteins, usually by cooking.
- Steam under pressure (with a grinding process) in large closed tanks.
- Cooks the material in its own fat by dry heat in open steam - jacketed drums.
- There covered proteins are used almost exclusively as animal foodstuffs, while there covered fats are used both industrially and in animal feeds.
- Considered an environmentally friendly method of disposal because it recycles the animal protein.
- Not economically feasible for poultry.
- The environmentally friendly concept.
- Production of a marketable product is the main advantages of rendering.
- Disadvantages include the fact that rendering is not always appropriate for disease situations because the carcasses must be transported to the plan.
- Some species are not amendable to efficient rendering practices.

Alkaline Hydrolysis

Alkaline hydrolysis or tissue digestion is a relatively new technique for carcass disposal. The process uses alkali at elevated temperature to convert the animal carcasses to a sterile aqueous solution of amino acids, sugars and soaps. The only byproduct of the process is the mineral constituents of the bone and teeth of the carcasses, which are soft enough after the organic matter, are degraded to be easily crushed by hand. The bone remnants can be captured and reused as calcium phosphate powder (sterile bone meal). The advantages of the process are that it sterilizes and digests in one operation, is more economical than some other forms of disposal, and is environmentally responsible.

All of the above methods have limitations for application in disaster situations, but because they are sometimes utilized in private or agricultural business settings as disposal methods, it is not inconceivable that they might be contemplated or discussed in disaster settings as viable options.

- Alkaline hydrolysis or tissue digestion is a relatively new technique for carcass disposal.
- The process uses alkali at elevated temperature to convert the animal carcasses to a sterile aqueous solution of amino acids, sugars and soaps.
- The only byproduct of the process is the mineral constituents of the bone and teeth of the carcasses, which are soft enough after the organic matter, are degraded to be easily crushed by hand.
- The bone remnants can be captured and reused as calcium phosphate powder S (sterile bone meal).
- The advantages of the process are that it sterilizes and digests in one operation, is more economical than some other forms of disposal.

□□□